BASIC MATH IN PLAIN ENGLISH

Bobby Rabon

authorHOUSE®

AuthorHouse™
1663 Liberty Drive
Bloomington, IN 47403
www.authorhouse.com
Phone: 1 (800) 839-8640

Published by AuthorHouse 03/24/2016

ISBN: 978-1-5049-7482-0 (sc)
ISBN: 978-1-5049-7481-3 (e)

Library of Congress Control Number: 2016901524

Print information available on the last page.

Introduction

Many students who have difficulty with math have a problem with the basic concepts, mainly because of conditioning. They spend much of their instructional time seeking answers to problems posed by the instructor or from the text. The answer usually consists of a number derived by use of a process, usually committed to memory, designed to find the answer without an understanding of the basic concepts involved. As a consequence, a habit of trying to find the answer becomes paramount—and all activity is directed toward that end. The conditioning is carried forth as an integral part of the math experience, and they immediately seek the answer before the proposed situation is evaluated. As a consequence, the focus is on the answer to the neglect of the evaluation necessary to arrive at a valid defensible conclusion.

This book is intended to diminish the immediate seeking of the answer and to magnify the idea of seeking to evaluate the situation to determine the information contained in the proposed situation (problem) before an attempt is made to find that elusive answer. To accomplish that aim, a thorough grounding in the basic fundamentals of the system is attempted by introducing the basic concepts without the stress of trying to immediately solve situations by applying memorized processes rather than gathering an understanding of the basic underlying concepts.

Basic Math in Plain English is an attempt to demystify math by comparing it to the basic structure of the spoken language, emphasizing the similarities, and stressing the differences. It is an attempt to show that mathematics is an integral part of our existence and how a basic understanding of the concepts that are paramount to the system of mathematics will create a greater appreciation of the role that math plays in our daily lives.

I implore the reader to carefully consider each salient idea and try to internalize its meaning for future reference. Algorithms (rules) are developed as shortcuts for arriving at defensible conclusions, but they do not necessarily enhance your understanding of the basis for the algorithms.

I have not taught mathematics for many years, but I was in administration and had many opportunities to interact with students. One of the basic questions was always if they were doing well in school. The answer was almost universal that math was the subject with which students were befuddled and could not understand or see any use for. The same questions posed to adults invariably elicit the same response. Why do intelligent persons have such difficulty with basic mathematics? Is math too difficult for the average person to comprehend—or is it presented in a manner that makes it incomprehensible to the average person?

I believe that the focus on using numbers without a clear definition of the nature of a number and how the numeration system is structured is one part of the problem. Another part of the problem is the focus on problem-solving techniques by getting answers rather than comprehending the structure and its use in constructing the algorithms commonly stressed in the teaching of math. If a teacher asks, "Why are two and two equal to four?" the answer is invariably, "It just is."

If the teacher asks, "What are the results of adding two and two?" the answer will be four. The four will be stressed since—from early in math instruction—students are conditioned to respond with the numeral four, but when asked if that is always true, the answer will always be yes. When asked to add two apples and two oranges, the answer will almost always be four fruits without the implied meaning of addition, which is often assumed and not explicitly expounded. Many other examples can be demonstrated, but we will not continue.

Mathematics is a tool like a hammer, telephone, or computer. Each tool has a function and can be used to attain certain objectives even without full comprehension of the structure and function. The structure and function of a simple tool such as a hammer is easy to decipher, but as tools become more complex, the function can be ascertained without a concurrent understanding of the structure. To attain a more complete understanding of the capabilities of the tool, both the structure and function need to be understood.

A telephone can be utilized without an understanding of how it functions or is constructed, but an understanding of the structure of a phone will create conditions for expanded use. The same is true for any constructed system. Mathematics is a constructed system, and a clear understanding of its structure and function will allow more utility in its use.

Imagine that you discovered some ancient coins and wished to decide the best use for them. Would they serve a better purpose if they were smelted into ingots or if they were kept intact? Now imagine that you spoke a different language. Would you come to the same conclusion? The same conclusion would be reached by the same person. The language would not change the reasoning process, hence the universality of the process. That process is devoid of numbers and is mathematical. The language of mathematics adds credence to the process.

This book is an attempt to mitigate some of the problems as discerned through interactions with students and other adults. It is hoped that by reading and rereading this book, some of the problems experienced with math will be tempered, the phobia associated with the word *math* will be proportionately diminished, and a rational discourse can be initiated by almost anyone who has a better understanding of the structure of the basic system of mathematical ideas and practices. This is an attempt to help readers understand the rationale for the current system of mathematics and how all the basic tenets of the system are explained by the basic model. Understanding the structure of the system will mitigate much of the confusion and apathy toward mathematics.

Exercises and problems to solve are not included in this book because the focus is directed toward the understanding of the basic structure and its uses in developing methods (algorithms) currently in use by most persons. A thorough scrutiny of the points of discussion will make it possible for the reader to develop algorithms (methods for solving problems) based on a complete understanding of the principles and not on rote memory without a clear understanding of mathematical sense.

From comments from students and others about math, most persons do not have a clear understanding of the function of mathematical reasoning and its implications to events in everyday life. There is little correlation between the relationships of mathematical ideas to situations in our daily lives. If "Do no act that will cause harm to others" is a deep and abiding principle of a society, it will help determine the actions of the citizens of that society. If the principle and its implied implications are not fully comprehended, then that will also impact the actions of the citizens. The same is true for the core principle (s) of mathematics, which if fully understood, would also affect the actions of the participants. One of the problems with math is that the basic core principles are not fully comprehended with the consequent result of math phobia, confusion, and apathy—any of which will cause a deficiency in comprehension.

Many problems in understanding basic math begin with the basic premise presented to beginning students. The fundamentals of math are assumed to be apparent when, in fact, they are not. An example is simple to illustrate. Most students are conditioned to make statements such as two plus two equals four, but the basis behind those statements is never completely understood. If one is asked, "What is two plus two?" the answer will invariably be four, and the student will be congratulated for that statement. Later, when we begin abstract algebra, when asked to add a plus b, we usually get a stutter and an incorrect response, usually ab. Why? A more profound question can be why one can add $1 + 2$, $3 + 4$, and so forth, with the concurrent explanation that we cannot add $a + b$? Another question should be how do the symbols 1, 2, 3 differ from the symbols a, b, c—and why do they act differently, if in fact they do?

If these questions are carefully examined, the discrepancies become a product of the lack of clarification of basic terms.

a. What is a number?
b. What is a numeral?
c. How are they different—and how can you distinguish one from the other?

The answer to similar simple questions will begin the process of removing some of the mystery that math imposes on many students. That will also help to answer several questions.

a. What are addition and subtraction?
b. What are numerals and what are numbers?
c. Why can one add $1 + 2$—but not $a + b$?
d. How is counting different from adding?
e. What are multiplication and division?
f. The importance and uses of zero and one in mathematics.
g. The understanding and use of *operations* in mathematics.
h. What is closure and why is it important?
i. Understanding the correct mathematical symbols and their uses.

Those are just a few of the problems with understanding basic math that will demystify some of the tenets of basic mathematical concepts. Many of the consistent problems encountered by math students can be mitigated by a thorough grounding in the basic, concise, workable definition of a number and how it is used to create the system of numbers in use today.

This book will mute many of the problems encountered by new math persons.

It is suggested that you create a dictionary of terms to be used for easy reference as needed.

Comparison of Math and English

Before we continue, a very important but little stressed point of mathematics is a concept called a truth table, which basically states that:

- you cannot deviate from the topic;
- only verified truths can be considered; opinions do not count as verified truths and cannot be considered;
    - Example 1: If the topic is about girls, then boys are not to be considered because they are not girls.
    - Example 2: If the topic is about students at school, both boys and girls can be considered because they are students.
- in general, if P is a statement and Q is another statement and ;
- if P is true and Q is true, then both P and Q are true (true conclusion);
    - Example: If P is a girl is true, and Q is a girl is also true, both P and Q are girls are also true.
- if P is true and Q is false, then both P and Q are not true (false conclusion);
    - Example: If P is a girl is true and Q is a girl is false, then both P and Q are girls are not true and the statement that both P and Q are not girls (boys) is also not true.
- if P is false and Q is true, then both P and Q are not true (false conclusion); or
    - Example: If P is a girl is false and Q is a girl is true, then both P and Q are girls is not true and the statement that both P and Q are not girls is also not true
- if P is false and Q is false, then both P and Q are not true (false conclusion).
    - Example: If P is a girl is false and Q is a girl is false, then both P and Q are girls is false.

The statements in are about girls, the subject is about girls, the conclusion must be about girls, the true conclusion must be about girls and they are a basis for the concept of addition. A very important concept of mathematics, illustrated by the above truth table, is that "one must stick to the subject."

An attempt will be made to flush out the verifiable truths in the following presentation of the structure and properties of the basic system of mathematics. Notice that this attempt is only a beginning and will be used to construct the system of mathematics.

Mathematics is a language—just as there are many different languages in use throughout the world today. The differences in languages are attributed to the differences in the original sounds (sounds of the alphabet) of which the language is composed. The different original sounds determine the sound of the combinations (words) that can be uttered using the basic sounds of the language. This can easily be corroborated by comparison of the sounds by use of the basic sounds of any language. An example can be illustrated by the use of the sounds in one language that cannot be produced using the basic sounds of another language, which partly determines the different words used to describe similar objects.

You may wish to experiment by using the alphabet of another language to try to produce certain sounds that are easily produced using your native language. If the sound of the alphabet in English B (bee) E (eee), T (tee) can be used to create the sound *bet,* and using the sound of those same alphabets in another language were B (bae), E (eae), T (tae), could the sound *bet* be produced? Notice that the words that can be produced are determined by the sounds of the alphabet. The languages of the world are different in part because of the differences in the original sounds of the native language that determines the words that can be produced using those sounds and the objects to which the sounds refer.

Notice also that there is a difference between the alphabet and the symbol that represents that alphabet. Can you distinguish the alphabet from the symbol that represents the alphabet? If asked to give me an A, and you yell out the sound of A and are given an A written on a piece of paper, one is the actual object and one is the symbol that represents that object. Notice also that both can be interchanged and used in communication of ideas, but the circumstances of their use determine which form will be used to convey the desired information. One can write information in a book or broadcast on the radio or other device. Notice that the medium determines the method.

If mathematics is treated as a language, it will also conform to the basic communication modes as shown above with at least one significant difference. In languages, words are assigned to objects, but in math, the object or symbol determines the word. In English, the word dog is assigned, and in German the word *Hund* is assigned to . In mathematics, the symbol 1 determines the name assigned to it in any language. In English, it is one, and in German, it is *ein.* It will be called by different names in different languages but the symbol will remain constant, hence the universal language.

There are basic differences between the language of mathematics and any language.

- Basic symbols and combinations can produce only one outcome in math, but they can produce more than one outcome in a spoken language.
- The combination *bat* can be assigned to more than one object in English. In math, only one symbol can represent any object (1, 2, 25, 1001, etc.).
- In languages, only a finite number of combinations can be produced without repetitions. In math, an unlimited number of combinations can be produced with no repetitions.
- In languages, combinations can have more than one meaning. In math, all combinations have only one meaning. (5 + 3) can have only one meaning, but *bat* can have more than one meaning or object to which it is assigned.

Mathematics is a language similar to any other language, and it must have some basic symbols that represent the alphabets of the system. Just as determined in spoken language, you will discern that a difference exists between the alphabet and the symbol that represents the alphabet. Can you determine the alphabets of math? Can you determine the symbol that represents each alphabet? If a numeral is the name of a number, and 2 is a numeral, then what represents the number 2? Conversely, if 2 represents a number, then what represents the numeral and are they one and the same thing? That is a relationship similar to a chair and a picture of a chair where the chair is the actual object and the picture is a representative of the chair but is not an actual chair.

To begin any system or project, one must have a basic plan or idea. Our basic plan will begin with symbols that are representative of groups (or members of a sequence) of items. The basic symbols (the alphabet of mathematics) are named 0, 1, 2, 3, 4, 5, 6, 7, 8, and 9. The symbols are called numerals. Each numeral represents a group (or unit of distance), and each group (or unit of distance) represents items in a sequence. Each numeral represents one group (one unit of distance) of items, which are numbers. Therefore, the numbers represented by the symbols (numerals) 0, 1, 2, 3, 4, 5, 6, 7, 8, and 9 are 1(0), 1(1), 1(2), 1(3), 1(4), 1(5), 1(6)1(7), 1(8), and 1(9). You will also notice that numbers and symbols (numerals) that represent them are different from each other. Numbers (actual item) consist of two parts, and numerals (name) consist of only one part. The names of the numbers are 0, 1, 2, and so forth, but the numbers are 1(0), 1(1), 1(2), and so forth.

A comparison of numbers and numerals is similar to the comparison of an object and a picture of the object. A picture can represent the actual object in certain circumstances, but the object must represent itself in other

Bobby Rabon

circumstances. Numbers and numerals operate in a similar manner. Numerals represent numbers in many applications—but not all as will be illustrated. Numerals can represent numbers in many applications, but not all.

If the symbols are placed side by side we get:
A. 0, 1, 2, 3, 4, 5, 6, 7, 8, 9, 0, 1, 2, 3, 4, 5, 6, 7, 8, 9, 0, 1, 2, 3, 4, 5, 6, 7, 8, 9, 0, 1, 2, 3, 4, 5, 6, 7, 8, 9 ….

Since these symbols can represent numbers as groups with increasing sizes (or unit distances with increasing distance from a beginning item), it becomes apparent that moving from left to right along the sequence above increases in value. From the above illustration, it is not possible to determine the value of any item upon casual inspection. There are only 10 alphabets that can be used to create an unlimited number of names of groups or positions. That can only be accomplished by naming groups with the same symbolism as the original groups were named. The groups become Group 0, because no complete group will have been used. When group 0 is completed group 1 will be used and the group number must be named in connection with the original alphabets and become 10, 11,12 and so forth. Groups of alphabets are listed to determine the value (distance from the origin) of the numeral or number.

                Group 0   Group 1
B. (0, 1, 2, 3, 4, 5, 6, 7, 8, 9)(10, 11, 12, 13, 14, 15, 16, 17, 18, 19),
                Group 2   Group 3 …….
(20, 21, 22, 23, 24, 25, 26, 27, 28, 29), (30, 31, 32, 33,3 4, 35, 36, 37, 38, 39).
                Group 10,   Group 11
(100,101,102,103,104,105,106,107,108,109)(110,111,112,113,114,115,116,117, 118,119)
                Group 12   Group 50
(120, 121, 122, 123, 124,125,126,127,128,129((500,501,502,503,504,505,50 6,507,508,509)
                Group 91   Group 92
(910,911,912,913,914,915,916,917,918,919) (920,921,922,923,924,925,926,9 27,928,929)

By use of the pattern in item B above, A becomes B and item B becomes item C below

C. 0,1,2,3,4,54,6,7,8,9,10,11,12,13,14,15,16,17,18,19,20,21,22,23,24,25,26,27, 28,29,30,31 and so forth and item B becomes item C below

8

D. 1(00), 1(01), 1(02), 1(03), <u>1(04), 1(05), 1(06), 1(07), 1(08), 1(09), 1(10),</u>
   <u>1(11), 1(12), 1(13), 1(14), 1(15)</u>.... 1(100), 1(109) 1(110), 1(111) ....1(119),
   1(120), 1(121) and so forth.

Item C is the expanded numerical sequence, and item D is the basic number sequence. Although D is the basic number line, the numeral line is more practical and will be used in most illustrations because the basic numeral line can represent the number line in most applications. Notice in C above that the numerals represent names of positions in the sequence. In D above, there is a counting part and a naming part. The number line D is the basic sequence to which all values are compared to get the answer and is referred to as the reference line. Every basic number contains the numeral 1.

Note that the representation of numbers in item D follows a pattern and the next item in the pattern is easily discerned. By following the same pattern as in D above, the system of whole numbers created.

Whole numbers are all numbers beginning with zero and continuing indefinitely in only one direction (to the right). Since all of the members of the sequence in C above have a numeral name, we say this is a *numerical sequence*. We can count by moving along the numerical sequence. Does the numerical sequence represent a number sequence? Absolutely, but does a numerical sequence act exactly the same as a number sequence? No, but it can be used to simplify calculations.

The placements represented by D above create a sequence that will be used as the basic number line, but the number line is represented by the numeral line in common practice. Note that any line could be used as a basic number line, but a definite position of the number line must also become part of the basic structure of the system. The basic (reference) position is called horizontal (east-west on earth). Notice that east-west movement implies no north-south movement and is moving in only one direction. That notion then implies that other basic number lines could also move in only one direction such as north-south on earth and would have no east-west movement. That notion implies that another basic number line could also exist that moves in only one direction (up-down on earth) that would have no east-west or north-south movement. Using this frame of thought, could there be other items or ideas that move in only one direction?

Three possible basic number lines can be created that move in only one direction from a reference point called zero. Each of these number lines is called a *dimension* that is spatial (dimensions of space). Dimensions are used as references for the construction of the basic system of mathematics.

The spatial dimensions are labeled x, (horizontal, east-west), y (vertical on paper but north-south on earth), and z (up-down on earth). This effort will deal primarily with the basic x dimension and to a lesser extent the basic y dimension. The dimensions x, y, and z are called axes.

Characteristics of Numbers

A number consists of two parts, (a counting part and a naming part). Each part of a number is a factor. The basic numbers are 1 (numeral): 1(5), 1(15), 1(105), 1(27). Basic numbers are binary because they consist of two parts.

Factor relates to the parts of a number. Every number consists of at least two factors. Some numbers are composites and can consist of more than two factors, but all basic numbers are written using only two factors, one of which is always one: 4 is a composite and can be written as 1(4) or 2(2), but the basic number 4 can only be written as 1(4) or 4(1).

Composites are numbers created by counting numbers and may be represented by more than one set of factors.

The number represented by 8 can be written using two sets of factors 1(8) and 4(2).

Prime numbers are numbers that can be written using only one set of factors:

1(7) is a basic number, but it can be represented with only two factors.

Power is a term used to indicate the number of times a factor appears in a number.

In the number 3(1) or 1(3), each factor appears only one time. The power of each factor is one, but as will be shown, all numbers with a power of zero have a value of one.

In the number 3(3), the factor repeats two times and is written 3^2. The raised two indicates that the factor 3 repeats itself two times in the composite number. The raised two is called an exponent. An exponent tells the number of times a factor appears in a number.

Moving along a numerical sequence is called *counting*, whereas moving along a number sequence is called *addition*. Notice that counting and addition are similar, but they are not exactly the same.

Counting is moving along a sequence of numerals. Numerals can be counted since they are increased by definite values on the basic sequence.

Counting numbers is called *addition*. *And* is the key word meaning "*to add.*" The addition symbol is plus (+). Addition is also a binary operation.

By definition, numbers are combined to form *composites* with the answer coming from the basic number line. The basic numbers increase by definite amounts; one part of the number is a counting part, and the other part is a naming part. The number 2(3) is a composite that is represented as 1(6) on the basic number line and can be found by counting $1(3) + 1(3) = (1 + 1)(3) = 2(3)$. *Remember that numbers that can be counted and given a common name can be added.* Remember also that numbers consists of two parts. If 1(2) and 1(2) is written as $1(2) + 1(2)$, $(1 + 1)(2) = 2(2) = 4(1)$ by counting $1 + 1$. Also notice that 2(2) can be written as $(1 + 1)(1 + 1)$ and is also equal to 4(1),

The number 3(5) can be written as $(2 + 1)(3 + 2)$ or $3(3 + 2)$ or $(5 – 2)(4 + 1)$, and many other ways. Can you determine why?

$$If\ 3(5) = (2 + 1)(3 + 2),$$
$$then\ (2 + 1)\ (3 + 2) = 2\ (3 + 2) + 1\ (3 + 2)$$
$$= 2\ (3) + 2\ (2) + 1\ (3) + 1\ (2)$$
$$= 6 + 4 + 3 + 2$$
$$= 15.$$

$$If\ 3\ (5) = (7 – 4)\ (3 + 2),$$
$$then\ (7 – 4)\ (3 + 2)\ = 7\ (3 + 2) – 4\ (3 + 2)$$
$$= 7\ (3) + 7\ (2) – 4\ (3) – 4\ (2)$$
$$= 21 + 14 – 12 – 8$$
$$= 15.$$

If the numbers were given different names, such as *a, b, c, d*, then the composites could look thusly.

If $1(x) = (a + b)$ and $1y = (c + d)$, the composite xy would appear as $(a + b)\ (c + d)$

$$If\ xy = (a + b)\ (c + d),$$
$$then\ (a + b)\ (c + d) = a\ (c + d) + b\ (c + d)$$
$$= ac + ad + bc + bd$$

$$If\ uv\ = (e – f)\ (g + h),$$
$$then\ (e – f)\ (g + h) = e\ (g + h) – f\ (g + h)$$
$$= eg + eh – fg – fh$$

$$If\ uv = (\ x + y)(x – y)$$
$$then\ (x + y)(x – y) = x^2 – xy + xy – y^2$$
$$= x^2 – y^2$$

Notice from the above examples that composites can be written as sums of numbers and are consistent with the definition of composite numbers.

Addition with Zero

Counting involves movement. Inherent in this statement is that counting by zero (0) does not change the value because there is no movement (0 can be written in many different forms).

$$0 = 2 - 2 \text{ or } -2 + 2$$
$$0 = 5 - 5 \text{ or } -5 + 5$$
$$0 = x - x \text{ or } -x + x$$
$$3 + 0 = 3 + (5 - 5)$$
$$3 - 0 = 3 - (7 - 7)$$

This idea can be used to change the appearance of a number without changing its value: 3 can be made to change its appearance by adding 0 as follows: $3 + (7 - 7) = (3 + 7) - 7 = (10 - 7) = 3$. Note also that in $3 - (7 - 7) = (3 - 7) + 7 = 3$ implies a definition for the symbol (−). Study the example and form an opinion based upon your observation. It will be discussed later.

$$3 + (5 - 5) = (3 + 5) - 5 = (8 - 5)$$

If x is a number, it can be made to change its appearance thusly by adding 0; therefore $x = x + (y - y) = (x + y) - y$ (note that only the appearance has been changed).

Additive Identity

Adding "0" to a number does not change its value; 0 can take many forms and can be used to change the appearance of a number without changing the value.

$$3(1) + 0(1) = (3 + 0)(1) = 3(1)$$
$$3 + (3 - 3) = 3 + 3 - 3 = 6 - 3 = 3(1)$$

Composite Numbers

Composite numbers are a group of numbers that are combined (counted) to make a new number (distributive property).

$$5(1) + 3(1) = (5 + 3)(1) = 8(1)$$

The equal symbol (=) means the same size or value.

Note that counting from left to right on a number line increases the value (distance from 0) of the number; counting from right to left on the number line decreases the value (distance from 0) of the number. Note also that a number consists of a counting part and a naming part. In the number 1(3), 1 is the counting part and 3 is the naming part. In the number 3(1), 3 is the counting part and 1 is the naming part. In the number 2 apples, 2 is the counting part and apples is the naming part.

Counting numbers produces a new number called a *composite number.*
1(3) + 1(3) = (1 + 1)(3) = 2(3); 2 is the counting part and 3 is the naming part; 1(6) is the position on the basic number line and is the answer.

3(4) + 2(4) = (3 + 2)(4) = 5(4) = 1(20) (distributive property)

ab + ab =1ab + 1ab = (1 + 1)(ab) = 2ab (ab is a composite with factors 1, a, and b.)

The "+" separates numbers and indicates that the counting part of the number is to be counted. The sign between numbers is called an *operation.*

Number Line Revisited
Fundamental Properties

B. 0, 1, 2, 3, 4, 5, 6, 7, 8, 9, 10, 11, 12, 13, 14, 15, 16, 17, 18, 19, 20, 21, 22, 23, 24, 25, 26, 27, 28, 29, 30, 31, 32 …

Before we continue, note that B above represents the basic numbers to which all other numbers are compared to get "the answer." Notice also that the sequence of numbers form a line called the *numeral line,* which is commonly referred to as the number line and is *one-dimensional.* The items on the number line increase or decrease by a definite (same) amount when moving along the line.

Addition

Moving (counting numbers) along the basic number line, note that the named part of the number must be the same for all objects counted. The subject cannot be changed. Conditionally, addition is moving to the right on the number line, and *subtraction* is moving to the left on the number line. Notice also that subtraction is a form of addition because they both involve movement on the number line. Subtraction reversed the direction of the anteceding (number after the sign) number. A negative (–) sign before a 3 means the direction of the 3 is reversed. In the statement 1(4) – 1(4), note that the second 4 is positive, but the negative sign before it reversed the direction and it moved in the opposite direction.

$$1(4) - 1(4) = (1 - 1)(4) = (0)(4) = 4(0) = 0$$

Reflexive Property

A very important concept in the language of math is the *reflexive property,* and it should be thoroughly understood and utilized.

Reflexive property means the order in which the parts of a number are written does not alter its meaning.

$$1(2) + 1(2) \text{ counted is } (1 + 1)(2) = 2(2) = 4(1)$$

$$2(1) + 2(1) = (2 + 2)(1) = 4(1)$$

B. 0, 1, 2, 3, 4, 5, 6, 7, 8, 9, 10, 11, 12, 13, 14, 15, 16, 17, 18, 19, 20, 21, 22, 23, 24, 25, 26, 27, 28, 29, 30, 31, 32 ....

In B and the two examples above, it is discerned that $1(2) = 2(1)$ because the group size must be the same for addition.

That statement illustrates an important and often underused mathematical concept called the *reflexive property.* This is a fundamental idea that can be used to describe other important basic ideas of the mathematical system.

Base sequence is the counting sequence using one (1) as the basic counting unit (sequence to which all others are compared).

Compare the sequences below with the basic sequence to discover if a relationship exists.

Can you find others and make predictions?

Sequence counting with 1 unit (Base sequence)

| 0 | 1 | 2 | 3 | 4 | 5 | 6 | 7 | 8 | 9 | 10 | 11 | 12 | 13 | 14 | 15 | 16 | 17 | 18 | 19 | 20 | 21 | 22 | 23 | 24 | 25 | 26 |

Sequence counting with 2 units

| 0 | | 2 | | 4 | | 6 | | 8 | | 10 | | 12 | | 14 | | 16 | | 18 | | 20 | | 22 | | 24 | | 26 |

Sequence counting with 3 units

| 0 | | | 3 | | | 6 | | | 9 | | | 12 | | | 15 | | | 18 | | | 21 | | | 24 |

Sequence counting with 4 units

| 0 | | | | 4 | | | | 8 | | | | 12 | | | | 16 | | | | 20 | | | | 24 |

Sequence counting with 5 units

| 0 | | | | | 5 | | | | | 10 | | | | | 15 | | | | | 20 | | | | | 25 |

Using the above, you can see that when all groupings are compared to the base group, the relationships becomes clear: $1(2) = 2(1)$; $1(3) = 3(1)$; $2(4) = 4(2)$ and so forth.

The reflexive property is very useful in math and should be very well understood and utilized. It allows us to add numbers that do not appear to be representing definite (the same) objects.

Add $1(2) + 2(1)$; note that they do not represent the same group size of items because one group contains two items, and one group contains one item and therefore cannot be added. The groups can be counted, but naming the composite is a problem. (This is an example of not changing the subject) But $1(2)$ can be written by the reflexive property as $2(1)$ without a change in value or $2(1) = 1(2)$.

$1(2) + 2(1) = 1(2) + 1(2) = (1 + 1)(2) = 2\ (2) = 4\ (1)$; or
$1(2) + 2(1) = 2(1) + 2(1) = (2 + 2)(1) = 4(1)$ either method gives the correct answer $4(1)$
$3(4) + 4(3) = 3(4) + 3(4) = 6(4)$ (by use of the reflexive property)
$4(3) + 4(3) = 4(3) + 4(3) = 8(3)$ (by use of the reflexive property)

Numerals are the names of numbers and numbers can be named with other symbols such as a, b, c, and so forth.

$a(b) = b(a)$ also when written as a composite
If $a = (c + d)$ and $b = (e + f)$, then $(a + b)(e + f) = (e + f)\ (a + b)$.

$2(a) = a(2)$

$1(a + b) = (a + b)(1)$
If one part of a number is written as a composite and the other as a basic number, the reflexive property will illustrate a new property called the *symmetric property*.

If $4(1) = 1(4)$ by the reflexive property and
$4(1) = 2(1) + 2(1)$, then (4 is a composite of $2(1) + 2(1)$).

$2(1) + 2(1) = 4(1)$ (symmetric property)

Symmetric property is a form of the reflexive property where both parts do not have the same appearance.

If $5(1) = 3(1) + 2(1)$, then $3(1) + 2(1) = 5(1)$
It is usually stated as: if $a = b$, then $b = a$ (introduces idea of equations).

$1(a + b) = (a + b)1$ is also symmetric.

Bobby Rabon

The symmetric property allows for the replacement (substitution) of items in a statement if it becomes necessary to complete an operation.

Add a + b if b = 3a; notice that a and b do not name the same quantity and cannot be added, but it is known that b = 3a; a + b can now be written as a + 3a = 4a by application of the symmetric property.

Commutative Property

Note that the reflexive property allows for numbers to be written in two distinct ways as illustrated above where 1(2) can be written as 2(1). This illustrates that parts of numbers can commute (change positions) without changing their value. Since addition is the counting of numbers, they can be counted in any order without a change in value. Changing positions is called *commuting.*

Commutative Property of Addition

The order in which numbers are written or added does not alter (change) the value. (Addition and multiplication only) Why? That statement refers to the fact that only additions can be commuted and deletions (subtractions) cannot be commuted. Addition and multiplication are basically the same operation of addition but subtraction and division are basically deletions and cannot be commuted and remain unchanged in value.

The order in which subtraction occurs does alter (change) its value. Review the definition of subtraction:

$3(2) - 4(2) = (3 - 4)(2)$ does not give the same outcome as $(4 - 3)(2)$.

The addition of numbers demonstrates another property of numbers called the *associative property.*

Associative Property of Addition

When adding numbers, they may be grouped in any convenient manner with no change in value. It also applies to multiplication. Why?

$3 + (4 + 5) = (3 + 4) + 5$

$3 + 9 = 7 + 5 = 12$

The order in which numbers are grouped when subtracting does change the value. Why? (Review the meaning of the "–" sign.)

Adding (counting) numbers to create composite numbers illustrates another property called the *distributive property*, which is the addition of numbers to form composite numbers.

$3(2) + 5(2) = (3 + 5)(2)$ (by symmetric property)
$(3 + 5)(2) = 3(2) + 5(2)$

Since a number consists of two parts, with one part for counting and the other part for naming, it follows that if Jim has one ant and Jan has one ant, then:

Example 1: 1 (ant) + 1 (ant) = (1 + 1) (ants) = 2 ant (s)
Example 2: 5 (cubes) + 3 (cubes) = (5 + 3) (cubes) = 8 cubes
Example 3: $ab + ac = (b + c)$ a or a $(b + c)$ (and by symmetric property
　　　　a $(b + c) = ab + ac$

Notice that the naming part of a number does not have to be a numeral and may be a composite.

One noteworthy idea from the examination of the basic number line is that for each operation of numbers on the number line, there is a unique (only one) "answer" that also lies on the number line. That is an example of a very important principle of numbers called the *closure principle*.

With $5(3) + 7(3) = 12(3) = 36(1)$, notice that 3, 5, 7, 12, and 36 all lie on the basic number line.

Closure Principle

The "answer" obtained from an operation with items from the system must also be a part of the system and creates a reliable system that allows for accurate predictions of events within the system and other similar systems.

Can you name another system that appears to be closed?
B. 0, 1, 2, 3, 4, 5, 6, 7, 8, 9, 10, 11, 12, 13, 14, 15, 16, 17, 18, 19, 20, 21, 22, 23, 24, 25, 26, 27, 28, 29, 30, 31, 32 …

Notice that from B above, if 1(3) is added to 1(3) we get $(1 + 1)(3) = 2(3)$, which is also on the number line as 1(6) or 6(1). This is one example of the closure principle that will later be used to predict the existence of negative numbers and other types of numbers as we develop the basis for the system of mathematics. Notice that the naming part of a number is enclosed in symbols of inclusion.

Symbols of inclusion are as shown (8), [8 (8)], {[6{7 (8)]} with the total grouping representing *one* number.

1(8), 1[8 (8)], and 1{6[7 (8)]}.

All entries contained within a symbol(s) of inclusion are treated as one number.

Remember that the symbols "+" and "–" separate numbers.

Example 1: [5 + (4 + 30)] + 10 represents only two numbers.

Example 2: [5 + (4 +30)] + 10 – {15 + [7 + (3 + 4)]} represents only three numbers.

Remember that the symbols *"+" and "–" separate numbers.*

Transitive Property

If two quantities have the same value, they are equal in value.

Example: If 3(1) + 4(1) = 7(1) and if 5(1) + 2(1) = 7(1), then 3(1) + 4(1) = 5(1) + 2(1). This allows one to replace objects if needed to complete an operation or solve a problem.

Example 2: add 2(6) + 12, but 12 = 2(6), therefore 2(6) + 12 = 2(6) + 2(6) = 4 (6).

Note that 12 was replaced by 2(6)y use of the transitive property.

Example 3: if a = b and if b = c, then a = c; that is the usual way the transitive property is written. Note that b was replaced by c.

This is a very important concept that allows for the replacement of items of equal value as needed. Replacements are sometimes called substitutions.

For Review

1. Adding numbers 5 (4) + 3 (4) = (5 + 3) (4) is known as the distributive property.
2. Writing numbers or composites by counting groups is known as multiplication.
3. Writing numbers or composites by subtracting groups is known as division.
4. The order in which a number is written does not change value, which is known as the reflexive property.

5. A number and its image may have different appearances, which is known as the symmetric property.
6. Quantities equal to a third quantity are equal to each other, which is known as the transitive property.
7. The order in which numbers are added or multiplied does not change their value, which is known as the commutative property.
8. The order in which numbers are grouped, when adding or multiplying, does not change their value, which is known as the associative property.
9. Two numbers whose product is always positive one are known as reciprocals.
10. Multiplication by one is known as multiplicative identity.

   Multiplication by one in its many forms can change the appearance of number without changing its value. One can look like any quantity multiplied by its reciprocal which is one divided by the number. If 1(2) is the number, then 1/2 is the reciprocal; if 1/2 is the number, then 2 is the reciprocal. Reciprocals will be discussed in detail later.

$$5 = 5(6)\frac{1}{6} = 5(1) = 5; \; a = a \, (a) \, \left(\frac{1}{a}\right) = a \, (1) = a$$

11. Multiplication by 0 means no movement from 0, and therefore $2(0) = 0$.
12. Division is the number of times one number can be subtracted from another number.
13. Division by zero is not defined and is not permitted.
14. Addition of zero to a number, which is known as additive identity, produces no change in value or movement. Addition of zero in its many forms can change the appearance of a number without changing its value.

$$5 + 0 = 5 + (5 - 5) = (5 + 5) - 5; \; x + 0 = x + (y - y) = (x + y) - y$$

Notice that all operations and statements refer to the basic symbols and the basic number line developed thus far. Many basic useful properties are listed above and can be applied in any order to solve problems when none are violated. This will be shown in greater detail in subsequent topics. Addition of zero to a number is known as *additive identity*.

Review the basic fundamental properties of the system of mathematics, which must be thoroughly understood to be usefully applied to situations involving mathematical operations.

Fundamental Ideas

1. Addition is counting numbers without changing their direction or value.
2. Subtraction is counting numbers with a change in their direction or value.
3. Counting is moving along the items named in a sequence of numerals.
4. Numeral is the name assigned to items on the number line.
5. A number line is the notation consisting of one and the numeral on the basic line.
6. Dimension is movement in only one direction from a reference point.

Symbols

Symbols of inclusion are symbols that indicate one item is included:

( ), [ ], { } can be grouped into multiples
[( )], { [( )] }, and any combination thereof.

"=" means that two quantities have the same value.
"+" means to continue without a change in direction.
"–" means to continue with a change of direction.

Definitions

Multiplication: Writing numbers or counting numbers to produce composites.
Division: Subtraction of equal-sized groups or multiplying factors of numbers by reciprocals of their factors.
Closure: The statement that nothing external to a system can be included in the system. All items and conclusions must be an integral part of the system.
Substitution: A form of the transitive property where one item can replace another item of equal value as needed to make a valid conclusion.
Binary: Basic numbers consist of two parts (one of which is always 1).

Sum: The result of adding is called the sum.
Product: the results of multiplication.
Substitution: Replacement of items of equal value in a statement.

Essential Properties

Reflexive property: the order in which a number is written does not change
    its value: $3(1) = 1(3)$ or $a(b) = b(a)$.
Symmetric property: A form of the reflexive property where the parts are
    not exactly alike. Since $3(4) = (2 + 1)(4)$, then $(2 + 1)(4) = 3(4)$
Distributive property: The result of counting (adding) numbers or
    decomposing numbers into corresponding parts:

If $1(3) + 2(3) = (1 + 2)(3)$, then $(1 + 2)(3) = 1(3) + 2(3)$
$6(7) = (4 + 2)(7) = 4(7) + 2(7)$

Transitive Property: If two different items have the same value as a third
item, the two different items have equal value
    If ten dimes equal one dollar and four quarters equal one dollar, then
ten dimes equal four quarters.

Commutative property of addition: The order in which numbers are added
does not alter their sum. Addition is binary.
Commutative property of multiplication: The order in which numbers are
multiplied does not alter the product.
Subtraction and division are not commutative.
Multiplicative identity is multiplication by one (1). Multiplication of a
number by one does not change its value. All basic numbers have one as
a factor.
The addition of zero (0) does not change the value.
Reciprocals are numbers whose product is always equal to positive one (1).

Other Definitions

Binary: Consists of two parts.
Factor: The parts of a number. All numbers are binary.
Power: The number of times a factor appears in a number.
Exponents: A number that indicates the number of times a factor appears
    in a number.

Number Line Revisited

Examination of the number line will show that addition on the number line involves movement from one place to another place on the line. Since moving along the number line is the same as counting and counting numbers is called addition, we can redefine addition as movement along the number line. *The numeral line is used to represent numbers.*

0, 1, 2, 3, 4, 5, 6, 7, 8, 9, 10, 11, 12, 13, 14, 15, 16, 17, 18, 19, 20, 21, 22, 23, 24, 25, 26, 27, 28, 29, 30, 31, 32, 33 …

Addition is movement along the number line. Moving from left to right increases the value of the number, and moving from right to left decreases the value of the number.

Movement can be in either direction on the number line and *cannot be restricted.*

Moving to the right is symbolized by use of a "+" (plus) sign and the word *and* or an equivalent.

Moving to the left is symbolized by use of a "–" (negative) sign and the word *difference* or an equivalent.

Notice also that the value of the number is dependent upon its position on the number line. Since movement is not restricted, movement can proceed to the left of the number zero on the number line. Since they must follow the same pattern as the positive numbers, which are *unique*, they are named negative numbers (opposite of positive numbers).

Zero is a reference point, and the value of all numbers is determined by the distance from the reference point (0). Since numbers are not the same as numerals and are represented as one group of a certain size or one unit of distance, they are written as 2 and 3, which are positive numerals, and (–2) and (–3), which are negative numerals. The number line now extends on both sides of the reference (zero) as illustrated below. Notice that numbers are written in symbols of inclusion.

20, 19, 18, 17, 16, 15, 14, 13, 12, 11, 10, 9, 8, 7, 6, 5, 4, 3, 2, 1, 0, 1, 2, 3, 4, 5, 6, 7, 8, 9, 10, 11, 12, 13, 14, 15, 16, 17, 18, 19, 20, 21

Notice that the numerals to the left of zero look exactly as the numerals to the right of zero, but they have different names and values because of their positions on the number line and must be unique. They are called negative numbers and are denoted by a "–" sign preceding the numeral.

−14, −13–12, −11, −10, −9, −8, −7, −6, −5, −4, −3, −2, −1, 0, 1, 2, 3, 4, 5, 6, 7, 8, 9, 10, 11, 12, 13, 14, 15, 16, 17, 18, ...

These numbers are called *integers,* which consist of positive and negative numbers.

Integers are positive and negative numbers as illustrated on the basic number line:

(−2) is a mirror image of 2, but it represents a different number and must be given a different name. However, they are the same distance from the reference point (0). Each number is represented by a point on the number line with each point exactly the same distance from the preceding point. Since a (−2) and 2 are the same distance from 0—the reference point—we say that the absolute distance from 0 is the same for both numbers. That distance is called the *absolute value* and is always represented as a positive number symbolized as | |. The symbol that indicates that absolute value is always positive. Note how the | | (absolute value sign) must be removed before an operation can be performed. (It is a symbol of inclusion.)

Absolute value is the distance of a number from the reference point (0) and is written thusly:

$$|2| = 2 \text{ or } |{-}2| = -(-2) = 2$$

The value must always be positive when the | | are removed, and they must be removed when calculations are made. Note that the removal of the | | from |−2| is accomplished by writing −(−2). Why? Refer to the meaning of a "−" sign before a number. If the number is given a different name such as a or (a + b), then three conditions must be considered when removing the absolute value sign prior to using in calculations:

Condition 1: If a represents a positive number, then $|a| = a$.
        If $a = 5$, then $|5| = 5$.
Condition 2: If a represents a negative number, then $|a| = -a$.
        If $a = -5$, then $|{-}5| = -(-5)$.
Condition 3: If a represents zero "0," then $|a| = 0$.
        If $a = 0$, then $|0| = 0$.

Since addition has been defined as movement along the number line, the symbols of operations are as follows:

A "+" sign between two numbers means to *continue* moving in the direction of the number and is called addition.

A "–" sign between two numbers means to *reverse* the direction of movement of the next number and is called subtraction.

(2) + (2) = (4). Note that the numerals are both positive and located to the right of zero.

(2) – (–2) = 4. Note that one numbers is positive and the other number is negative, but the direction of the second number is reversed.

(2) – (2) = (0). Note that both the numbers are positive but the direction of the second number is reversed.

(–2) – (–2) = 0. Note that both numbers are negative but the direction of the second number is reversed.

(2) + (–2) = 0. Note that one is positive and the other is negative, but the direction of the second number is not reversed.

(–3) – (–3) = 0; (-3) + (3) = 0; (–3) + (–3) = –6; (3) + (3) = 6

Noteworthy: The numbers to the left of 0 on the number line must be used in the same manner as numbers to the right of 0 on the number line.

Example 1: since (2) + (2) = 4, then (–2) + (–2) = –4.

Example 2: since (2) – (2) = 0, then (2) + (–2) = 0.

Example 3: since (2) + (–2) = 0, and (2) – (2) = 0 then (2) – (2) = (2) + (–2) by use of the transitive property.

This illustrates a new definition of subtraction.

Subtraction is the addition of the opposite of the subtrahend (second number).

(5) – (5) = 5 + (-5); x – y = x + (-y); (p + q) – w = (p + q) + (-w)

Counting numbers is called addition (remember that addition is counting numbers with the same named parts), which produces composite numbers. Numbers consist of two parts, and the writing of a number or a composite is aptly called multiplication.

1(2) + 1(2) = (1 + 1)(2)= 2(2)

Notice that counting numbers produces a new number, which is called a *composite*. Since numbers have the same form as composites, they represent the same operation, which is called *multiplication* indicated by the word *of.*

Multiplication: the writing of a number or adding numbers to form a composite number.

5(1) + 5(1) = (5 + 5)(1) = 10(1) or 1(10)

When a numeral is written as a number, it is called a product.
The numeral 2 written as a number 1(2) produces a product: 3(2) is a product. The number 6(1) or 1(6) is also a product

Multiplication Numbers in Our System

2(3) = 6 means 1(3) + 1(3) or 3(1 + 1) = 2(3) (the product)

2 (–3) = –6 and 3(–2) = –6 because 1(–3) + 1(–3) = (1 + 1)(–3) = 2(–3)

What if –2 multiplies –3? What will be the magnitude of the product? (Review the definition of subtraction.)

–2(–3) = –1(–3) + [– 1 (–3)] Note that the "–" sign before a number reverses the direction of that number.

Therefore, –1(–3) = +1(3) and –1(–3) –1 (–3) = 1(3) + 1(3) = (1 + 1)(3) =2 (3) = +6. Notice the grouping symbols the meaning of the – sign.

–2(–3) = 6; –3(–5) = 15; –5(–6) = 30

What if –2 multiplies +3? What will be the magnitude of the product? (Review subtraction.)

–2 (3) = 3(–2) by use of the reflexive property
–2 = (– 1) + (–1) by using the definition of addition
3(–2) = 3[(–1) + (–1)] by use of substitution for the value of –2
3(–2) = –1 (3) + [–1(3)] by use of the meaning of + sign
3(–2) = –6

When multiplying two factors—if the factors are opposite in direction, the product will be negative.

When multiplying two factors, if the factors are the same in direction, the product will be positive.
Can you make a deduction about multiplying three factors?

7 – 7 = 7(1) – 7(1) = 7(1 – 1) = 7 (0) = 0 or 1(7 – 7) =7(0) = 0

Notice that the – means subtraction and the second seven is positive. (Review writing numbers.)

The sign between numbers denotes the operation of addition or subtraction.

The direction of a number will be positive unless otherwise stated to be negative by enclosure in symbol of inclusion (parentheses).

Example 1: in $7 - 7$, both numerals are considered positive, and the $-$ sign indicates the operation of subtraction is to be performed: $7 - 7 = 7(1) - 7(1) = 7(1 - 1) = 0$ or $(1 - 1)(7) = 0$. All numbers are enclosed in symbols of inclusion, but for simplicity, the symbols of inclusion are usually omitted for the naming part of positive numbers. $1(2)$ can be written simply as 2, but -2 is written as (-2).

Example 2: In $7 - (-7)$, the sign between the numbers indicates the operation of subtraction is to be performed and the second number is a negative number; therefore,

$7 - (-7) = 14.$

Division of Numbers

If multiplication is adding numbers, then subtracting numbers, which is a form of addition, cannot be called multiplication and will be called *division*. In division, the number from which subtraction is taken is called the *dividend*, the number subtracted is called the *divisor*, and the number of times subtraction occurs is called the *quotient*. In the example below, 8 is the dividend, 2 is the divisor, and 4 is the quotient.

Division is counting the number of times a smaller number can be subtracted from a larger number.

Example 1: 2 can be subtracted from 8 exactly 4 times. $8/2 = 4$

	Number of times 2 is subtracted
$8 - 2 = 6$	1
$6 - 2 = 4$	2
$4 - 2 = 2$	3
$2 - 2 = 0$	4

Notice that 2 is larger than 0, and we must stop. Therefore, 2 can be subtracted from 8 exactly 4 times. Therefore $\underline{8} = 4$
2

Example 2: 4 can be subtracted from 12 exactly 3 times. 12/4 = 3

	Number of times 4 is subtracted
12 – 4 = 8	1
8 – 4 = 4	2
4 – 4 = 0	3

Example 3: Using the idea expressed above, divide 6786 by 6 (or the number of times 6 can be subtracted from 6786).

Note that 6 can be subtracted from 6786, 1000 times and 1000 (6) = 6000.

6786 – 6000 = 786	1000

6 can be subtracted from 786, 100 times and 100 (6) =600

786 – 600 = 186	0100

6 can be subtracted from 186, 31 times and 31 (6) = 186

186 – 186 = 0	0031
Total number of times 6 can be subtracted from 6786:	1131

That same example could have been done thusly:

Note that 6 can be subtracted from 6786, 1000 times;1000 (6) = 6000   1000
    6786 – 6000 = 786

Note that 6 can be subtracted from 786, 100 times and 100 (6) = 600
    0786 – 600 = 186   0100

Note that 6 can be subtracted from 186, 20 times and 20 (6) = 120
    186 – 120 = 66   0020

Note that 6 can be subtracted from 66, 11 times and 11 (6) = 66
    66 – 66 = 0   0011

Total number of times 6 can be subtracted from 6786   1131

    That same example could have been done thusly:

Note that 6 can be subtracted from 6786, 1000 times and 1000 (6) = 6000
    6786 – 6000 = 786   1000

Note that 6 can be subtracted from 786, 100 times and 100 (6) = 600
    786 – 600 = 186   0100

Note that 6 can be subtracted from 186, 10 times and 20 (6) = 120
    186 – 120 = 66   0020

Note that 6 can be subtracted from 66, 10 times and 10 (6) = 60

$66 - 60 = 6$ 0010

Note that 6 can be subtracted from 6, 1 time and 1 (6) = 6

$6 - 6 = 0$ 0001

Total number of times 6 can be subtracted from 6786:   1131

$\dfrac{6786}{6} = 1131$ $\dfrac{6568}{25} = 262 + \dfrac{16}{25}$ $\dfrac{16497}{351} = 47$

```
       1131                    262                         47
6000| 6786             2500|6568               3510|16497
      6000                   5000                     14040
  600|786               250|1568               351| | 2457
      600                    1500                     2457
   60|186                  25|68                      0000
      180                    50
     6|6                     16
       6
       0
```

Example 4: 4 can be subtracted from 17 four times with some excess: 17/4 = 4 with an excess of 1. Since 4 is larger than 1, and if subtraction is stopped, fractions are created.

The extra part of a group that contains four parts and is written as $\frac{1}{4}$, and $\frac{1}{4}$ is a fraction because it does not represent a complete set.

Fractions have the form *of a/b, where a and b are integers*. Fractions are sometimes referred to as ratios.

Divide 8 by 1 written as (8/1)   Number of times 1 can be subtracted
Number of times 1 can be subtracted

$8 - 1 = 7$	1
$7 - 1 = 6$	2
$6 - 1 = 5$	3
$5 - 1 = 4$	4
$4 - 1 = 3$	5
$3 - 1 = 2$	6
$2 - 1 = 1$	7
$1 - 1 = 0$	8

Division of a number by one does not cause a change in the output.

Division of a Number with Zero as a Denominator

How many times can the number (0) be subtracted from any number?
　　Example 5: Suppose 4/0 (by definition of division).

$4 - 0 = 4$	1
$4 - 0 = 4$	2
$4 - 0 = 4$	3

Zero can be subtracted from any number an unlimited number of times. Therefore we can safely say that division by zero is not defined and is not permitted. No denominator of a fraction can be equal to zero. Can you use the definition of subtraction to explain why this is true?

Example: $\frac{(x+y)(x-y)}{xy}$, neither x nor y can be equal to 0 because 0 (y) = 0 and 0 (x) = 0.

Example: $\frac{x+5}{x+3}$; x cannot have a value of –3 because the divisor would be equal to 0.

Rational and Irrational Numbers

The closure principle states that fractions must also be part of the system of numbers on the number line. Therefore, the number line now must consist of numbers that are positive, negative, and fractions. The new name for the system is *rational numbers*. Notice that rational numbers consist of whole numbers, integers, and fractions. The parts of a fraction are separated by a bar (horizontal line between the parts of the fraction) as is shown in the fractions $\frac{a}{b}, \frac{x+y}{p+q}$ and $\frac{3+3a+5b}{6a+7b}$.
The *bar* is a symbol of inclusion and represents one number.

In a fraction, the top number is called the *numerator.* The bottom number is called the *denominator.* The denominator describes the number of parts contained in the whole object; the numerator tells the number of parts of the object that are to be used or considered useful for a particular situation.

　　Notice from the above that multiplication is adding groups of equal size to form a composite, and division is subtracting groups of equal size to form a quotient.
　　The statement 5(6) means to count groups of equal size (6) to form a composite number called 30, which is 1(30).

The statement 35/7 means to subtract groups of equal size (7) and counting the number of times it can be done, which is five times.

A number consists of two parts; one part counts the groups, and the other part is for the size or name of the group.

5(4), 7(3), 20(350), a(b) or as commonly written ab, 2/3, a/b.

Irrational Numbers

All rational numbers can be written as fractions, but there are some numbers that cannot be written as fractions. Some of these numbers are written with fractional exponents such as $2^{1/2}$, $3^{1/2}$, and $5^{2/3}$.

The numbers that cannot be written as fractions are called *irrational numbers*.

Rational numbers can be written as fractions or as decimals that have sequences that repeat in a pattern that may or may not terminate. Rational numbers as decimals with repeating patterns are represented as fractions.

Example 1: 1/3 = 0.333333 is rational (repeating unending pattern)
Example 2: 1/4 = 0.750 (terminating pattern)
Example 3: 1/12 = 0.0833333 … (a unending, repeating pattern)
Example 4: 7/9 = 0.77777…. (a unending, repeating pattern)

Irrational numbers cannot be written as fractions or decimals with repeating patterns.

Example 1: $2^{1/2}$ = 1.414213562 (with no definite pattern and no end)
Example 2: $3^{1/2}$ = 1.732050808 (with no definite pattern and no end)
Example 3: $5^{1/2}$ = 2.236067978 (with no definite pattern and no end)

There are no integers with a value of $2^{1/2}$, $3^{1/2}$, $5^{1/2}$, $10^{2/3}$ and many others, but there are some integers with fractional exponents hat can be written as fractions. Some examples are $4^{1/2}$, $9^{1/2,}$ $64^{1/2}$, $81^{1/2}$, $169^{1/2}$. Study the examples to seek a possible explanation.

The group (set) of numbers that include both rational and irrational numbers are called *real numbers*. Real numbers include all familiar numbers, but they do not include all possible numbers. Review the closure principle for a possible clarification. Those other numbers can be introduced (but not at this juncture).

Real numbers include all types of numbers discussed thus far including:

- whole numbers
- integers
- rational numbers (including all numbers that can be written as fractions)
- irrational numbers (numbers that cannot be written as fractions)

Fractions as Decimals

Division creates fractions, and fractions can be written in a form called decimals when the subtraction process is continued by the addition of a point to separate the whole part from the fractional part and continuing as shown below.

Example 1: Since 27/5 means the number of times 5 can be subtracted from 27 with two parts of the group of 5 left over written as 5 + 2/5. A point is placed after the subtraction of the smaller number from the larger number, and the process is continued by adding zeros and continuing to subtract as illustrated below:

Number of times 5 is subtracted

$27 - 5 = 22$	first
$22 - 5 = 17$	second
$17 - 5 = 12$	third
$12 - 5 = 7$	fourth
$7 - 5 = 2$	fifth written as 5 + 2/5; 2/5 means two parts of a

group containing five objects.

Continue by adding a point *(decimal)* to separate the parts as shown below to create decimals. Note that 2/5 deals with parts of a whole that consists of five parts and we added zero parts. The statement 5 + 2/5 can now be written as 5 + 2.0/.5. Note that when 2.0 was considered, 0.5 had to be considered to keep our numbers in the proper place. We know that 5 can be subtracted from 20 four times and since that is true it follows that 0.5 can be subtracted from 2.0 a total of 0.4 times; 5 + 2/5 can now be written as 5.4 in decimal form with .4 = 2/5.

Notice that subtraction created division, division created fractions, and fractions can be written as decimals.

−14, −13, −12, −11, −10, −9, −8, −7, −6, −5, −4, −3, −2, −1, 0, 1, 2, 3, 4, 5, 6, 7, 8, 9, 10, 11, 12, 13, 14, 15, 16, 17, 18

The number line now includes the whole numbers, integers, and fractions as determined by the closure principle. The system that includes whole numbers, integers, and fractions is aptly called rational numbers because they can all be written as fractions. Whole numbers can be written as fractions because, as shown above, division of a number by one does not change the number. Rational numbers can also be written as decimals. Can all fractions be written as decimals? Take a guess, and we will explore that proposition later.

Adding Fractions

Reciprocals and Cancellation

Since fractions are numbers, consist of two parts (as do all numbers), and adding numbers is done by counting groups of like size, fractions are to be combined as any other numbers.

Example 1: $1 \left(\frac{1}{2}\right) + 1 \left(\frac{1}{2}\right) = (1 + 1) \left(\frac{1}{2}\right) = 2 \left(\frac{1}{2}\right) = 1$

Note: The division bar is written as / on the computer.

Example 2: $2 \left(\frac{1}{3}\right) + 5 \left(\frac{1}{3}\right) = (2 + 5) \left(\frac{1}{3}\right) = 7 \left(\frac{1}{3}\right) = \frac{7}{3}$

Example 3: $5 \left(\frac{1}{12}\right) + 6 \left(\frac{1}{12}\right) = (5 + 6) \left(\frac{1}{12}\right) = 11 \left(\frac{1}{12}\right) = \frac{11}{12}$

Example 4: $\frac{1(3)}{2(4)} + \frac{5(3)}{2(4)} = \frac{3}{8} + \frac{15}{8} = \frac{18}{8} = \frac{2(9)}{2(4)} = \frac{9}{4}$

Example 5: $2 \left(\frac{a}{b}\right) - 3 \left(\frac{a}{b}\right) = (2 - 3)\frac{a}{b} = -1 \left(\frac{a}{b}\right) \text{ or } -\frac{a}{b}$

Example 6: $\frac{(a + b)}{d} c + \frac{(a + b)}{d} c = 1c \frac{(a + b)}{d} + 1c \frac{(a + b)}{d} = 2c \frac{(a + b)}{d}$

When adding fractions, if the number of the counting parts (numerator) is the same as the number of the naming parts (denominator), they are called *reciprocals*. In the statement $(1 + 1 + 1) / (1 + 1 + 1) = 3/3$, the counting three parts is the same as the naming parts. When that the case, they are called reciprocals. Notice that $\frac{3}{3}$ can be written as $3 \left(\frac{1}{3}\right)$.

3 can be subtracted from 3 exactly 1 time; the reciprocal of 3 is 1/3; a is 1/a; 2ab is 1/2ab.

Reciprocals are numbers whose product is always equal to 1 or a number with a quotient of 1.

Example 1: $5 \left(\frac{1}{5}\right) = \frac{5}{5} = 1$; $8 \left(\frac{1}{8}\right) = \frac{8}{8} = 1$; $a \left(\frac{1}{a}\right) = \frac{a}{a} = 1$; $(a + b) (1)/a + b) = \frac{(a + b)}{(a + b)} = 1$

Reciprocals are very useful mathematical tools that can be used to transform dissimilar items into items with similar names as illustrated below.

Add the fractions that do not have the same name or represent the same quantity

Example 1: $\frac{1}{2} + \frac{1}{3} = \frac{1(3)}{2(3)} + \frac{1(2)}{3(2)} = 3\left(\frac{1}{6}\right) + 2\left(\frac{1}{6}\right) = (3 + 2)\left(\frac{1}{6}\right) = \frac{5}{6}$

Example 2: $\frac{3}{4} + \frac{5}{6} = \frac{3(6)}{4(6)} + \frac{5(4)}{6(4)} = 18\left(\frac{1}{24}\right) + 20\left(\frac{1}{24}\right) = (18 + 20)\left(\frac{1}{24}\right) = \frac{38}{24} = \frac{2(19)}{2(12)} = \frac{19}{12}$

Note that the reciprocal was used three times in this example.

Example 3: $\frac{a}{b} + \frac{c}{d} = \frac{a(d)}{b(d)} + \frac{c(b)}{d(b)} = \left(\frac{1}{(bd)}\right)(ad + bc) = \frac{ad + bc}{bd}$

Note that each fraction was multiplied by the reciprocal of the other fraction, and they were transformed into fractions with the same name or value of (1/bd).

Reciprocals can also be used in a process called *cancellation* by writing one of the factors of a number as reciprocals (as shown below). This is called reducing *to lowest terms*.

$$\frac{38}{24} = \frac{19(2)}{12(2)} = \frac{19}{12}, \quad \frac{25}{75} = \frac{25(1)}{25(3)} = \frac{1}{3}, \quad \frac{98}{114} = \frac{2(49)}{2(57)} = \frac{49}{57}$$

$$\frac{304}{2304} = \frac{2(152)}{2(1152)} = \frac{2(2)(76)}{2(2)(576)} = \frac{2(2)(2)(38)}{2(2)(2)(288)} = \frac{2(2)(2)(2)(19)}{2(2)(2)(2)(144)} = \frac{19}{144}$$

Cancellation occurs when the numerator and the denominator of a fraction are reciprocals.

Cancellation with the use of the multiplicative identity and reciprocals is a form of division,

Divide 1/2 by 1/2	Divide 3/4 by 1/2	Divide 5/6 by 2/3

$$\frac{\frac{1(2)}{2(1)}}{\frac{1(2)}{2(1)}} = \frac{1}{1} = 1$$

$$\frac{\frac{3(2)}{4(1)}}{\frac{1(2)}{2(1)}} = \frac{6}{4} = \frac{6}{1} \cdot \frac{1}{4}$$

$$\frac{\frac{5(3)}{6(2)}}{\frac{2(3)}{3(2)}} = \frac{15}{12} = \frac{15}{1} \cdot \frac{1}{12}$$

Use of Reciprocals in Problem Solving

Reciprocals can be used to change the appearance of quantity without changing the value.

Since 2.2 pounds are equivalent to 1 kilogram, reciprocals can be used to change the appearance of an outcome without changing the value of the outcome.

If 2.2 lbs. = 1 kg, then by multiplying both members by the reciprocal of 2.2 lbs. as shown

$$\frac{2.2 \text{ lbs.} (1)}{2.2 \text{ lbs.}} = \frac{1 \text{ kg} (1)}{2.2 \text{ lbs.}} \text{ we will see that } 1 = \frac{(1 \text{ kg})}{2.2 \text{ lbs.}}$$

If 2.2 lbs. = 1 kg is multiplied by the reciprocal of 1 kg
1 kg = 2.2 lbs., multiply by the reciprocal of 1 kg and get
$1 \text{ kg}\frac{(1)}{1kg} = \frac{2.2 \text{ lbs.}}{1 \text{ kg}}$, we get the statement $1 = \frac{2.2 \text{ lbs.}}{1 \text{ kg}}$.

By using the idea expressed above any number's appearance can be changed by the use of the reciprocal to produce a quantity whose value is one (1).

Example 1: How many kilograms are in sixty pounds?

$$60 \text{ lbs.} = (60 \text{ lbs.}) \frac{(1 \text{ kg})}{(2.2 \text{ lbs.})} = \frac{60 \text{ kg}}{2.2} = 27.28 \text{ kg}$$

Notice that lbs/lbs cancel and become 1.

Example 2: Change 1 mile to yards if 1 yard = 3 feet and 1 mile = 5280 feet.

$$1 \text{ mile} = (1 \text{ mile}) \frac{(5280 \text{ feet})}{(1 \text{ mile})} \frac{(1 \text{ yard})}{(3 \text{ feet})} = \frac{5280 \text{ (yard)}}{3(1)(1)} = 1760 \text{ yard (s)} \quad \text{miles and feet cancel to 1}$$

Notice that the factor one was used two times in the above statement, which implies that it can be used multiple times if necessary.

Example 3: How many centimeters are in a foot if 1 in. = 2.54 cm and 1 foot = 12 in.

$$1 \text{ foot} = 1 \text{ foot} \frac{(12 \text{ in.})}{(1 \text{ foot})} \frac{(2.54 \text{ cm})}{(1 \text{ in})} = (12 \text{ in})\frac{(2.54\text{cm})(1)(1)}{(1)(1)} = 30.48 \text{ cm}$$

Notice that feet/feet cancel to 1 and inch/inch cancel to 1

Review of Concepts: Definitions

Absolute Value: The distance that a number is from the reference point (0) symbolized by | |.
Absolute Value: represented by the distance from "0" and always has a positive value.
Addition Symbol "+" (and): To continue moving in the same direction as the original direction.
Adding Fractions: Use of reciprocals to transform the denominators to the same name or value and apply the definition addition
Addition: Counting, is movement on the number line, of the non-numerical part of numbers
Additive Identity: Adding zero to a number causes no movement and does not change the value of the number.
Associative Property: When adding numbers, they may be grouped in any convenient manner as needed to make computations.
Basic Number Line: the number line from which the answer comes: 2(2) = 4(1) or 1(4)
Basic Number: consists of two parts with a numerical part and 1: 1(0), 1(2), 1(9) (written enclosed in symbols of inclusion). All numbers are considered positive unless stated otherwise by the inclusion of a negative sign: (–2) and not as –2 because the 2 is considered positive.

Parts of all numbers are enclosed in symbols of inclusion as illustrated: 1(2), 5(6), 9(19). Numbers with other names are usually not enclosed in a symbol of inclusion: ab, xy, x (a + b) and so forth but all numbers contain "1" as a factor.
All numbers that are not positive are written as
1(–2), 5(–4), but the reflexive property allows for
1(–2) = –2(1) or 2(–1).

Basic Symbols: 0, 1, 2, 3, 4, 5, 6, 7, 8, 9.
Cancellation: a condition where the numerator and denominator have a common factor(s)

Bobby Rabon

Closure Principle: All problems and answers must be contained within the system of mathematics. Closure allows for the making of accurate predictions of other facets this closed system as well as other closed systems.

Commutative Property: The order in which a number is written or added does not alter (change) its value.

Composite Number: the result of counting (adding) numbers.

Counting: Moving along the numerical sequence.

Decimal: Writing fractions without a denominator: $3/4 = 0.75$.

Digits: The parts of numbers such as 1(2) where 1 and 2 are digits.

Dimension: A sequence that moves in only one direction from a reference point; a basic number line.

Distributive Property: Addition of numbers to form composite numbers.

Division by Zero: Division by zero is not permitted by the definition of division.

Division by One: Division by one does not change the value of a number and is permitted.

Division: The number of times one number can be subtracted from another number.

Exponent: The raised numeral that tells the number of times a factor appears in a number.

Factor: The parts of a number.

Integers: Positive and negative numbers increasing or decreasing by unit values.

Multiplication: The writing of a number or a composite number.

Multiplicative Identity: Multiplication by one does not change the value of a number.

Negative Numbers: Numbers to the left of zero, predicted by the closure principle.

Number Line: the arrangement of the numerals into a sequence increasing or decreasing by unit values.

Numeral: The names assigned to the basic symbols.

Power: The number of times a factor appears in a number.

Prime Number: A number that has only two factors. Some numbers can have more than two factors.

Product: The result of multiplication (the answer from the basic number line).

Rational Number: Any number that can be written as a fraction or as a decimal that terminates or has a repeating pattern of non-ending digits.

Reciprocal: The result obtained when the factors of a number produces a product of one.

Reference Line: The basic number line from which all answers come

Reference Point: The starting point on the number line called zero (0) or origin.

Reflexive Property: The parts of a number can be interchanged without changing the value.

Subtraction Symbol "–" (difference): To move in the opposite direction from the original direction.

Symmetric Property: Property where each member has a different appearance but same value.

Transitive Property: When each of two quantities is equal to a third quantity, the two quantities are equal to each other.

Whole Numbers: Numbers that begin a 0 and continue in successive increments to create other groups using the basic symbols as a guide.

Writing of Numbers: Numbers are written with grouping symbols ( ), [ ], { } or any combination thereof.

Rational Numbers: Numbers that can be written in the form of a/b where a and b are integers or written as decimals that produce non-ending or repeating patterns when written as decimals.

Irrational Numbers: Numbers that are not rational and cannot be written as a fraction or a non-ending or a repeating sequence when written as a decimal

+ Sign: Separates numbers, and means to continue in the same indicated direction.

2(3) + 2(–3); note that all factors are positive except –3; the + sign between the numbers means the – 3 continues in the same direction with the result.

$2(3) + 2(–3) = 1(6) + 1(–6) = 0$

$2 (3) + 2 (3) = 1 (6) + 1 (6) = 12$ (in neither example did the direction change)

– Sign: Separates numbers and means to reverse the direction of the number after the sign.

$2(3) – 2(3) = 6 – 6 = 0$; note that all factors are positive, but the negative sign reversed the direction of the number after the sign.

$2(3) - 2(-3) = 6 + 6 = 12$; note that all factors are positive except the –3, but the direction was reversed by the "– "sign that separates them.

$2(3) - 2(3) = 6 - 6 = 0$; the "–" sign between them reversed the direction.

Real Numbers: All numbers shown thus far, including both rational and irrational numbers

Powers, Factors, and Exponents

Basic numbers are binary (consist of two parts): a counting part and a naming part called factors. Numbers that have all factors represented by the same symbol can be written by indicating the symbol with a small raised number called an *exponent*. The exponent tells the number of times a factor is used, and it is the same as addition and must obey similar rules as is done when adding numbers. Additionally, the use of exponents makes for easier representations and compilations. Some examples are listed below.

The number called 4 can be written as 4(1) or 2(2) or 2^2.
The number called 9 can be written as 9(1) or 3(3) or 3^2.
The number called 16 can be written as 16(1) or 2(8) or 4(4) or 4^2.
The number called 25 can be written as 25(1) or 5(5) or 5^2.
The number called 36 can be written as 36(1) or 18(2) or 4(9) or $6(6) = 6^2$.

Each example cited above is an example of a square number (square). Notice that the number of groups and the name of the group are represented by the same symbol. A number represented by three same symbols is called a cube number (cube); those represented by four same symbols and above are simply called *powers*.

Factor: The parts of a number or composite number.
Power: The number of times a factor appears in a number. The number of times is represented by a raised number is called an exponent.
Example of factor: 1(2) (The factors are 1 and 2.)
Example of a power: 1(2) (The power of factor 1 is 0; the power of factor 2 is 1.)
Example of a power: 2(2) (The power of each factor is 1, and the power of the number is the sum of the exponents and is written $2^{(1+1)} = 2^2$.)
Example of a power: 4(4) (The power of each factor is 1, and the power of the number is 2 written $4^{(1+1)} = 4^2$.)

The power of a number with three common factors is 3; 4 common factors is 4, and so forth:

$3(3)(3) = 3^3, 3(3)(3)(3) = 3^4$

The raised number that represents the common factors is called the *exponent* and is raised above the affected number, which is called a *base*.

When multiplying numbers with common powers, the exponents obey the same rules as the original numbers for the operations of addition and subtraction. Why?

Since multiplication is a form of addition and division is a form of subtraction, to multiply powers, exponents are added and to divide powers, they are subtracted.

Example 1: $2 (2) = 2^1 (2^1) = 2^{(1+1)} = 2^2$

Example 2: $4^1 (4^1) (4^1) = 4^{(1+1+1)} = 4^{(3)}$

Example 3: $(25^2)^3 = 25^{(2+2+2)} = 25^6$ which is the same as multiplying the powers $3(2)$

Example 4: $5^{-1} (5^{-1}) = 5^{[-1+(-1)]} = 5^{-2}$

Example 5: $5^1 (5^{-1}) = 5^{(1+(-1))} = 5^{(1-1)} = 5^0$

Example 6: $5^1/5^1 = 5^{(1)-(1)} = 5^0 = 1$

Example 7: $5^4/5^4 = 5^{(4-(4))} = 5^0 = 1$

Example 8: $5^4/5^{-2} = 5^{[4-(-2)]} = 5^6$ (remember the rule for subtraction)

Examination of examples 5 and 6 will show:

$5^1 (5^{-1}) = 5^{(1-1)} = 5^0 = 1$; $5 (1/5) = 1$ and by the transitive property $5^1 (5^{-1}) = 5 (1/5) = \dfrac{5}{5}$

Note: any number raised to a power of zero is equal to one.

Example 9: $x^0 = 1$; $100^0 = 1$

Examination of examples 6 and 7 indicate that the numeral named *1 has a power of 0.*

A number that has two factors that are exactly the same is a square. Four has two factors that are exactly the same and represents a square. Numbers that have factors that are the same are called *perfect*.

$4 = 2(2), 9 = 3(3), 16 = 4(4), 25 = 5(5), 36 = 6(6), 49 = 7(7), 1 = 1(1), 64 = 8(8)$, and so forth

5^2 is a perfect square: 5^3 is a perfect cube: 5^4 is a perfect power.

From the examples listed above, 2(2), 3(3), 4(4), 5(5), 6(6), 7(7) are all square numbers or squares, and all squares are composites. They are also perfect squares.

Example 1: $2(2) = 4$, $2(2) = 2^{(1)}(2)^{(1)} = 2^{(1+1)} = 2^{(2)} = 4$. Note that 2 is the base. Example 1 states that to multiply powers you must add exponents
Example 2: $3^{(2)3} = 3^2 (3)^2 (3)^2 = (3)^{(2+2+2)} = (3)^6$, also $3^{(2)3} = 3^{(6)}$
Example 2 states that to raise a power to a power exponents are multiplied.

If the factors of a number are not exactly the same, then the number does not represent a perfect power.

If a number has three factors that are exactly the same, the number represents a perfect cube.

$1 = 1(1)(1)$, $8 = 2^1(2^1)(2^1)$, $27 = 3^1(3^1)(3^1)$, $64 = 4(4)(4)$, $125 = 5(5)(5)$, $216 = 6(6)(6)$, $343 = 7(7)(7)$
    Each of the same factors has a power of one. The power of the number is the sum of the powers of the factors.
    First power: the number contains one factor other than one. Remember that any number raised to a power of zero is one.
    Second power: the number contains two factors that are exactly the same: 11(11).
    Third power: the number contains three factors that are exactly the same: 9(9)(9).
    Fourth power: the number contains four factors that are exactly the same: 5(5)(5)(5).
    Power: factors of a number are called powers.
    Powers are indicated by a raised number above a factor (base) that indicates the numbers of factors represented called an exponent.

2(1) has only one factor other than one.

2(2) has two factors that are the same: (base) ----> $2^{2\,(exponent)}$.

2(2)(2) has three factors are the same: (base) ----> $2^{3\,(exponent)}$.

2(2)(2)(2) has four factors are the same: (base) ----> $2^{4\,(exponent)}$.

x (x) (x) (x) (x) (x) (x) has seven factors are the same: x^7.

(a + b)(a + b) has two factors that are the same: (base) $(a + b)^2$
All numbers have one as a factor with a power of zero.

Base: The factors that repeat.
Power: The raised number that tells how many times the factor (base) repeats.
Exponent: Another name for a power.
Co-efficient: the counting part of the number that contains the base $3x^2$
       (3 is the co-efficient with a power of 1)
$4^3 = 4 (4) (4)$: 4 is the base, 3 is the power (exponent) and 1 is the co-efficient
$3(5^2) = 3(5)(5)$: 5 is the base, 2 is the power (exponent) and 3 is the co-efficient
$w(9^5) = w(9)(9)(9)(9) (9)$: 9 is the base, 5 is the power (exponent) and w is the co-efficient
$(x + y)^4$ $(x + y)$ is the base and 5 is the power (exponent).

Can numbers other than whole numbers be raised to a power?

Example 1: $(1/2)^2 = \dfrac{(1) (1)}{(2) (2)} = \dfrac{(1)^2}{(2)^2} = \dfrac{1}{4}$

Example 2: $(2/3)^2 = \dfrac{(2) (2)}{(3) (3)} = \dfrac{(2)^2}{(3)^2} = \dfrac{4}{9}$

Example 3: $(2/3)^3(2/3)^2 = \dfrac{[(2)(2)(2)][(2)(2)]}{[(3)(3)(3)][(3)(3)]} = \dfrac{(2)^{(3 + 2)}}{(3)^{(3 + 2)}} = \dfrac{(2)^5}{(3)^5} = \dfrac{32}{243}$

Since powers of numbers are increased by adding exponents, it follows that in division the powers are reduced by subtracting exponents.

$\dfrac{8}{2} = \dfrac{2^3}{2^1} = 2^{(3 - 1)} = 2^2$ $\dfrac{64}{16} = \dfrac{4^3}{4^2} = 4^{(3 - 2)} = 4^1$; $\dfrac{40}{4} = \dfrac{5(8)}{4} = \dfrac{5(2^3)}{2^2} = 5(2)^{(3 - 2)} = 10$

By use of the definition of numbers and reciprocals, we may write

$\dfrac{8}{2} = 8 \dfrac{(1)}{(2)}$ $\dfrac{64}{16} = 64 \dfrac{(1)}{(16)}$ $\dfrac{40}{4} = 40 \dfrac{(1)}{4}$

You will notice that division can be represented as multiplication and must obey the same conditions as multiplication.
     You should also remember that division involves subtraction of groups (sets).
     Since multiplication of powers involves adding exponents, the division of powers must involve the subtraction of exponents.

When you divide with factors that are the same, you must subtract exponents.

$\dfrac{20}{4} = \dfrac{4^1\,(5)}{4^1\,1} = 4^{(1-1)}\,(5) = 4^{(0)}\,(5) = 1(5),$ factors 4 cancel each other and the factors 5(1) remain.

Notice that cancellation is the same as division and cancellation of common factors always leaves one as a factor. Therefore cancellation is:

$$\dfrac{7^1}{7^1} = 7^{(1-1)} = 7^0 = 1 \qquad \dfrac{25^1}{25^1} = 25^{(1-1)} = 25^0 = 1 \qquad \dfrac{25^1}{25^0} = 25^{(1-0)} = 25$$

$$\dfrac{5^3}{5^3} = 5^{(3-3)} = 5^0 = 1; \qquad \dfrac{25^4}{25^4} = 25^{(4-4)} = 25^0 = 1$$

Any factor that has zero as an exponent has a value of one.

$\dfrac{1000}{1000} = 1$ and $\dfrac{1000^1}{1000^1} = 1000^{(1-1)} = 1000^0$, then $1000^0 = 1$ (review symmetric property).

If $\dfrac{x^7}{x^7} = x^{(7-7)} = x^0$ and $\dfrac{x^7}{x^7} = 1$, then $x^0 = 1$

What happens when an exponent has a value of less than zero?

$\dfrac{2^2}{2^3} = \dfrac{2\,(2)\,(1)}{2\,(2)\,(2)} = 2^{(2-3)} = 2^{(-1)}$ and $\dfrac{2^2}{2^3} = \dfrac{2\,(2)\,(1)}{2\,(2)\,(2)} = \dfrac{1}{2}$ by cancellation. Therefore, $2^{-1} = \dfrac{1}{2}$

$\dfrac{1}{2} = \dfrac{2^0}{2^1} = 2^{(0-1)} = 2^{-1.}$ Therefore, $2^{-1} = \dfrac{1}{2}$

The symmetric property states that if $\dfrac{1}{2} = 2^{(-1)}$, then $2^{(-1)} = \dfrac{1}{2}$.

$\dfrac{2^3}{2^7} = \dfrac{2(2)(2)(1)(1)(1)(1)}{2(2)(2)(2)(2)(2)(2)(} = \dfrac{1(1)(1)(1)}{2(2)(2)(2)} = \dfrac{1}{2^4}$ also $\dfrac{2^3}{2^7} = 2^{(3-7)} = 2^{(-4)}$ Therefore $2^{-4} = \dfrac{1}{2^4}$

If $\dfrac{2^3}{2^7} = 2^{-4}$ and also $\dfrac{2^3}{2^7} = \dfrac{1}{2^4}$, then $\dfrac{1}{2^4} = 2^{-4}$. Review negative exponents and symmetric property.

Note: Any number with a negative exponent can be written as a fraction.

$3^{(-2)} = \dfrac{1}{3^2} \quad a^{(-b)} = \dfrac{1}{a^b}.$ The symmetric property indicates that

$\dfrac{1}{3^2} = 3^{(-2)}$ and $\dfrac{1}{a^b} = a^{(-b)}.$

Operations with Powers

Addition of Powers

$2^3 + 2^3$ means $1(2^3) + 1(2^3) = (1 + 1)(2^3) = 2(2^3) = 2(8) = 16$

$4^3 + 4^3$ means $1(4^3) + 1(4^3) = (1 + 1)(4^3) = 2(4^3) = 2(64) = 128$

$5^3 + 6^3$ means $1(5^3) + 1(6^3) = (1 + 1)?$ It is unclear by what the question mark is to be replaced

$7^3 + 6^3 = 1 (7^3) + 1 (6^3) = (1 + 1) (?)$ It is unclear by what the question mark is to be replaced.

$7^3 + 7^4 = 1 (7^3) + 1 (7^4) = (1 + 1) (?)$ It is unclear by what the question mark is to be replaced.

$7^3 - 7^4 = 1 (7^3) - 1 (7^4) = (1 - 1) (?)$ It is unclear by what the question mark is to be replaced.

Notice that both the base and the exponents must be the same when adding powers of numbers.

Multiplication of Powers

$2^1 (2^1)$ means $2(2) = 4$ It also means $2^{(1 + 1)}$ or $2^2 = 4$

$2^2 (2^3)$ means $4(8) = 32$ It also means $2^{[(1 + 1)+ (1+1+1)]}$ or $2^{(2 + 3)} = 2^5 = 32$

$2^3 (2^4)$ means $8(16) = 128$ It also means $2^{[(1+ 1+ 1+ 1)+ (1+ 1+ 1+ 1)]}$ or $2^{(3 + 4)} = 2^7 = 128$

$5^2 (5^4)$ means $25 (625) = 15625$ It also means $5^{(2 + 4)} = 5^6 = 15625$

$2^3 (3)^2$ means $2(2)(2)(3)(3) = 8(9) = 72$ It does not mean $2(2)(2)(2)(2)$ or $2^5 = 32$ nor does it mean $3(3)(3)(3)(3)$ or $3^5 = 243$

$3^2(4^2)$ means $9(16) = 144$ It does not mean $3(3 (3)(3)$ or $3^4 = 81$ nor does it mean $4(4)(4) (4)$ or $3^4 = 256$

The power of a number is equal to the sum of the power of its factors

Raising Powers to Powers

If $10 = (5 + 5)$, then $(10)^2 = (5 + 5)^2$.

If $25 = (16 + 9)$, then $(25)^3 = (16 + 9)^3$.

If $y = x^2$, then $y^2 = (x^2)^2$.

Bobby Rabon

If $64 = 8^2$, then $64^2 = (8^2)^2 = 4096$ or
$$64^2 = 8^2 \, (8^2) = 64 \, (64) = 4096 \text{ or}$$
$$64^2 = 8^{(2+2)} = 8^4 = 4096.$$

If $36 = 6^2$ then $36^2 = (6^2)^2 = 6^4$.

If $y = x^2$, then $y^3 = (x^2)^3 = x^{2\,(3)} = x^6$.

What is the rule for raising powers to powers? Refer to the addition of basic numbers before making conclusions.

When you multiply with factors that are the same, you add exponents.

When you multiply powers of the same factors, you multiply exponents.

Division of Exponents

By use of the definitions of division and factoring, we may write:

$$\frac{8}{2} = 4(2)(\underline{1}) = 4 \qquad \frac{15}{5} = 3(5)(\underline{1}) \qquad \frac{35}{7} = 5(7)(\underline{1}) \qquad \frac{120}{12} = 120(\underline{1}).$$

You will notice that division may be represented as multiplication by the reciprocal of the denominator and must obey the same conditions as multiplication.

In $\frac{8}{2} = 2(2)(2)(\underline{1}) = 2(2)$, a factor 2 in the numerator is canceled by a factor $\frac{1}{2}$ (reciprocal of 2).

In the denominator (use of reciprocals where $2(1/2) = 1$), multiplication by the reciprocal of a number is called *cancellation* because the product is always 1.

Notice that there are three factors of 2 in the numerator 8 and one factor of 2 in the denominator, and $\frac{8}{2} = \frac{2(2)(2)(1)}{1(1)(1)(2)} = 2(2)(2)(\underline{1}) = 2^2$ can also be written as $\frac{8}{2} = \frac{2^3}{2^1} = 2^{(3-1)} = 2^2$

$$\frac{32}{8} = \frac{2(2)(2)(2)(2)}{1(1)(2)(2)(2)} = 2(2) = 2^2 \text{ can also be written as } \frac{32}{8} = \frac{2^5}{2^3} = 2^{(5-3)} = 2^2$$

Notice that when you divide with factors that are the same, you may subtract exponents. (Review subtraction of numbers.)

Notice that when you multiply with factors that are the same, you may add exponents.

44

When you divide with factors that are the same, you may subtract exponents or cancel with principle of multiplication by the use of reciprocals.

$$\frac{20}{4} = \frac{4(5)}{4(1)} = 5 \qquad\qquad \frac{35}{5} = \frac{7(5)}{1(5)} = 7$$

The factors 4 cancel each other, and the factors 5 (1) remain. Why?

Suppose the exact same factors are in both the numerator and the denominator of a fraction as listed below:

$$\frac{5(5)(5)(5)(5)(5)(5)(5)}{5\,(5)\,(5)\,(5)5\,(5)5\,(5)} = 1 \text{ or } \frac{5^8}{5^8} = 5^{(8-8)} = 5^0 = 1$$

$5^0 = 1;\ 15^0 = 1;\ 150^0 = 1$

What can you conclude about x^0?

What happens if a number is raised to a power that is less than zero but is an integer? Powers tell how many times a factor can be used as a base.

A base can be any number, including fractions and other types of numbers to be introduced at a later time. The powers can be any numbers, including fractions.

Study the sequence below to see if a pattern can be discerned.

-2^5	-2^4	-2^3	-2^2	-2^1	2^0	0	2^0	2^1	2^2	2^3	2^4	2^5
-32	16	-8	4	-2	1	0	1	2	4	8	16	32

Study the sequence below to see if a pattern can be discerned. Compare the patterns and note the differences and similarities, if any exists.

-2^{-5}	-2^{-4}	-2^{-3}	-2^{-2}	-2^{-1}	-2^0	0	2^0	2^1	2^2	2^3	2^4	2^5
1/-32	1/16	1/-8	1/4	1/-2	1	0	1	2	4	8	16	32

What can you conclude from the patterns listed above?

Some observations to be concluded from are:

Positive numbers raised to a negative power are represented as fractions

$$5^{-2} = \frac{1}{25}$$

Negative numbers raised to positive power are represented as positive numbers if the powers are even numbers (0, 2, 4, 6, 8.) and negative numbers if the negative powers are odd (–1, –3, –5).

Examples: $-5^2 = 25 \quad -5^3 = -125 \quad -5^{(4)} = 625 \quad -5^{(7)} = -78125$

Bobby Rabon

Negative numbers raised to a negative power are represented as positive fractions if the powers are even (–2, –4, –6...) and negative fractions if the negative powers are not even (–1, –3, –5).

Examples: $(-3)^{-2} = \dfrac{1}{(-3)^2} = \dfrac{1}{9}$ $(-3)^{-3} = \dfrac{1}{(-3)^3} = \dfrac{1}{-27}$

Negative numbers raised to a negative power are represented as positive fractions if the powers are even (–2, –4, –6...) and negative fractions if the negative powers are not even (–1, –3, –5).

The examples below illustrate the basic fundamental idea of the fact that numbers are binary and that the + sign means the direction of movement is unchanged and the – sign means that the direction is reversed.

Example 1: $(-3)^2 = -3(-3)$. Note that the number consists of 2 parts (binary) and the negative sign means that the direction is reversed hence $(-3)^2 = -3(-3) = 9$.

Example 2: $(-3)^3 = -3(-3)(-3)$. Since $-3(-3) = 9$ (the direction is changed), then $9(-3) = -27$ since the direction is not changed because of the $9(-3)$.

Fractional Powers of Numbers

Quantities can also be raised to fractional powers called *roots.*

Since numbers raised to a power of zero always equals one, then numbers raised to a fractional power will always be greater than one. A square root is a number raised to a power of 1/2. A cube root of a number is that raised to a power of 1/3. A fourth root is a number raised to a power of 1/4.

Those are examples of using prior knowledge to gain new insights into problem solving.

Notice from illustrations above that $2^0 = 1$ and $2^1 = 2$, then $2^{1/2}$ must lie between 1 and 2 on the number line and must be related by the following statement.

$2^0 < 2^{1/2} < 2^1$ where < means "is less than."

$1 < 2^{1/2} < 2$

$2^{1/2}$ lies between 1 and 2 and is written as $\sqrt{2}$ (square root of 2), which means you must find a number that can be raised to a power of 2 and has a value of 2. The number must be bigger than one and smaller than two.

Similarly $3^0 = 1$ and $3^1 = 3$; and $3^{1/2}$ must lie between 1 and 3 or $3^0 < 3^{1/2} < 3^1$; $3^{1/2}$ is the square root of 3 written as $\sqrt{3}$. Notice that a number raised to the power of two must equal three. That is $(3^{1/2})^2 = 3$. Because $2^2 = 4$ and $3 < 4$, $3^{(1/2)}$ must be bigger than 1 and smaller than 2.

Find the value of $5^{1/2}$.

$5^0 = 1$ and $5^1 = 5$; $5^{1/2}$ must lie between 1 and 5. Why?
It is known that $2^2 = 4$ and $4 < 5$, also $3^2 = 9$ and $5 < 9$. It is also known that the square root of 4 is 2 and the square root of 9 is 3 and since $4 < 5 < 9$, then $4^{1/2} < 5^{1/2} < 9^{1/2}$. It is now evident that $2 < 5^{1/2} < 3$.
$5^{1/2}$ must be represented by a number between 2 and 3 on the number line.
You can use your calculator to find the $5^{1/2}$, but can you find it without the calculator?

Find the value of $7^{1/2}$.

If $7^0 = 1$ and $7^1 = 7$; $7^{1/2}$ must lie between 1 and 7 or $7^0 < 7^{1/2} < 7^1$, since 4 is less than 7 and 7 is less than 9, and $4^{1/2} = 2$ and $9^{1/2} = 3$ it follows that $7^{1/2}$ must be more than 2 but less than 3 written symbolically as $2 < 7^{1/2} < 3$.
The answer by can be found with trial and error by finding the square of numbers between two and three to see if you can find a value close to three when the number is squared.

1. Try $(2.5)^2 = 6.25$
2. Try $(2.6)^2 = 6.76$
3. Try $(2.63)^2 = 6.9169$
4. Try $(2.634)^2 = 6.937956$
5. Try $(2.63456)^2 = 6.940906394$. $7^{1/2} = 2.63456$ approximately. A definite conclusion cannot be obtained. Why?

Suppose a number is raised to a power other than 1/2. If 4 is raised to the $2/3$ power, then it is the cube root of the square of 4 and is written as $\sqrt[3]{4^2}$, $a_{b/c} = \sqrt[c]{a^b}$. Remember that numbers can be given any name and can be given the names of a, b, and c (if necessary).

Examination of the numbering system will reveal that the numbers we considered until fractional powers were included could be written as fractions (one number divided by another number even if the divisor is 1).

Example:

$\frac{5}{1}$	$\frac{7}{8}$	$\frac{55}{13}$	$\frac{a}{b}$	$\frac{x+y}{3x+4y}$	$\frac{50+51}{75}$

Examination of fractional powers will show that not all numbers can be written as fractions and are not rational numbers. *Rational numbers include all numbers that can be written as a fraction.* Numbers exist that cannot be written as fractions and are not rational numbers are *irrational numbers.*

Suppose we write fractions and their decimal equivalents as listed below. (Use a calculator.)

1/3 = 0.33333….
1/2 = 0.50000 …
11/12 = 0.9166666 …
19/36 = 0.527777…
7/9 = 0.7777…
13/18 = 0.72222…
8/13 = 0.615384615384…..
31/27 = 1.148148148…

If you continue the division, can you predict what the next numeral will be? Each of these fractions and their decimal equivalents has something in common. Can you determine what? Look carefully for a pattern.

If you continue the calculations with fractions, can you predict what the next numeral will be?

1/11 = 0.09090909…
1/9 = 0.111111…
2/3 = 0.6666666…
7/13 =0.538461538…
25/36 = .0694444…
21/33 = .063636363…
13/21 = 0.619047619…
35/44 = .795454545…

If you continue the calculations with fractional powers, can you always predict what the next numeral will be?

Use your calculator and find the value of $2^{1/2}$, $3^{1/2}$, and $5^{1/2}$.

Examine the decimal places and compare what is seen with the results when fractions are written as decimals using your calculator.

$2^{1/2}$ =1.414213562...
$3^{1/2}$ = 1.732050808...
$5^{1/2}$ = 2.236067978...
$11^{1/2}$ = 3.31662479...

If you continue the calculations with fractions, can you predict what the next numeral will be?

When you compare the fractions written as decimals and the fractional powers written as decimals, are they similar or different? How are they similar and/or different? Review the definition of rational numbers and reconsider your answer if needed.

If fractional exponents are revisited and roots of numbers are taken, some can be written as fractions and some cannot be written as fractions. Numbers that cannot be written as fractions are *irrational* (not rational) numbers. Some irrational numbers are $2^{1/2}$, $3^{1/2}$, and $5^{1/3}$.

Equations

Equations are statements that quantities have the same value but do not necessarily look alike. This is indicated by the symmetric property, which is written as "If a = b, then b = a." Notice that the statements represent the same items, but they do not look alike. Each side of an equation is called a *member*. In an equation x = y, x is a member and y is also a member.

True Equation: a statement where quantities have the exact same appearance and value.

Example: The numeral 3 has the same appearance as 3, or 3 = 3, 10 = 10, 101 = 101, etc.

Note: A number can be given any name we choose if we do not know its exact position on the number line

Variable Equation: an equation in which one or both members contain a number whose position on the number line is not exactly known.

Example: 2x = 16, 3a = 2a + 4, etc.

Truth Table implies that all statements must be true for the conclusion to be true.

The number line will be represented by the *numeral line* for efficiency.

>Example 1: If 2 is on the number line and 5 is on the number line, the sum must also be on the number line. The sum must be unique (only one). All statements must be true for the conclusion to be true.
>
>Example 2: $2 + 3 = 5$; 2 is true, 3 is true, and 5 is also true.
>
>Example 3: If $2(3) + 4 = 10$, then $2(3) + 4 + 5 = 10 + 5$ is still true and remains an equation.
>
>Example 4: If $2(3) + 4 = 10$ is true, then $2(3) + 4 - 6 = 10 - 6$ is still true and is an equation
>
>Example 5: If $5 = 5$ is a true equation, then $5 + 3 = 5$ is not a true equation. If $5 = 5$ is a true equation, then $5 = 5 + 3$ is not an equation.
>
>Example 6: If $x = 25$ is an equation, then $5x = 5(25)$ is still an equation.
>If $x = 25$ is an equation, $\dfrac{x}{5} = \dfrac{25}{5}$ is still an equation.

The examples above will demonstrate that equations can be manipulated without ill effects if the same operation is performed equally on both members. That is an ironclad property that cannot ever be violated.

Notice that the solution of the equation is an application of the properties of numbers discussed earlier, and they can be applied in any order as long as they are not violated. You may wish to solve the examples in other ways without violating any of the basic properties of the number system studied earlier.

Example 1: Solve: $9x + 10 + 5x + 4 = 12x - 18$

$9x + 10 + 5x + 14 = 12x - 18$ (Combine all similar items in each member)

$(9x + 5x) + (10 + 14) + 18 = 12x - 18 + 18$ (Use grouping and equation properties)

$14x + 42 = 12x$ (Combine all similar items in each member)

$(14x - 12x) + 42 - 42 = 12x - 12x - 42$ (Use equation properties and grouping.)

$2x = -42$ (Combine all similar items in each member)

$\frac{1}{2}(2x) = \frac{1}{2}(-42)$ (Use reciprocals to find the value of x.)

$x = -21$ (Simplify by use of reciprocals)

Check to see if each equation is true by substituting in the original equation.

Example 2: Solve $2x + 5 = 11$ by using methods 1 and 2 below.
1. $2x + 5 = 11$ by subtracting 5 from both members.
2. $2x + 5 = 11$ by multiplying both members by the reciprocal of 2.

Example 3: Solve $2x + 5 = 11$.

$2x + 5 = 11$ (Find the value of the unknown number x.)

$2x + 5 - 5 = 11 - 5$ (Subtract 5 from both members and combine similar items.)

$2x = 6$ (Simplified.)

$1/2(2)x = 1/2(6)$ (Multiply both members by the reciprocal of 2.)

$x = 3$ (Unknown number found.)

The same problem can be solved in a different way(s):

Example 4: $2x + 5 = 11$ (Find the unknown number by using a different method.)

$\frac{1}{2}(2x + 5) = \frac{1}{2}(11)$ (Multiply both members by the reciprocal of 2.)

$x + 5/2 = 11/2$ (Use distributive property to separate x from the group.)

$x + 5/2 - 5/2 = 11/2 - 5/2$ (Simplify by subtracting 5/2 from both members.)

$x = 6/2$ Definition of addition (counting numbers $1/2(11) - 1/2(5)$

$x = 3$ (Use factoring and reciprocals or definition of division.)

Which method do you like best?

Use your chosen method to solve $3x - 5 = 13$.

In each process, no properties of the system were violated—and the solution was the same for both procedures. This illustrates how the solution of an equation can be modified by the person as long as no tenets of the system are violated. The method of solving a problem is dependent upon the insight of the solver and is not set by any rule except to remain true to the system and always be faithful to the developed concepts. Notice also that solutions must be obtained by using known properties of the system such as reciprocals, distributive, etc.

Which method would you prefer? Notice that each solution did not violate any of the properties of math that were developed. (It seems that a better choice for solving equations would be to combine terms by addition or subtraction before performing any other operation)

Each side of an equation is called a member. The equation is maintained if the same operation is performed on each member.

Solving Equations

Any action that is performed on one member of an equation must be performed on the other member of the equation to avoid changing the meaning of the statement to make it an untrue statement. Remember that each side of the equation is called a member. All equations must become true equations to be considered solved.

Example 1:   If $2x + 3 = 11$ is a variable equation, then
$$2x + 3 - 3 = 11 - 3$$
$$2x = 8$$
$$2(1/2)x = 1/2(8)$$
$$x = 4$$

$x = 4$ a variable equation where the value of x is known.
Replace the x in $2x + 3 = 11$ and get $2(4) + 3 = 11$, which produces a true equation $11 = 11$.

Example 2: If $5x - 3 = 2x + 6$ is a variable equation, then
$$5x - 3 + 3 = 2x + 6 + 3, \text{ which becomes } 5x = 2x + 9$$
$$5x - 2x = 2x + 9 - 2x, \text{ which becomes } 3x = 9$$
$$1/3(3x) = 1/3(9)$$
$$x = 3$$

$x = 3$ where the value of the variable x is known.
Replace x with 3 in the original statement to determine if it is a true equation.

Note: The above statements are true because of addition.

Example 3: $3x - 6 = 9$ Try to solve this equation with at least two different approaches.

Solving More Complex Equations

Solving equations involves maintaining the truth of the original statement by performing the same operation(s) on both members of the equation to maintain the truth of the statement.

Example: Solve: $3x + \dfrac{1}{5} = x + \dfrac{3}{5}$

$$5(3x - \tfrac{1}{5}) = 5(x + \tfrac{3}{5}) \text{ (Multiply both member by the reciprocal 1/5.)}$$

$15x - 1 = 5x + 3$ (Simplify by use of distributive property.)

$15x - 1 - 3 = 5x + 3 - 3$ (Subtract 3 from both members.)

$15x - 4 = 5x$ (Simplify by adding like terms.)

$15x - 5x - 4 = 5x - 5x$ (Subtract 5x from both members.)

$10x - 4 = 0$ (Simplify by adding like terms.)

$10x - 4 + 4 = 0 + 4$ (Add 4 to both members.)

$10x = 4$ (Simplified.)

$1/10(10x) = 1/10 (4)$ (Multiply both members by the reciprocal of 10.)

$x = 4/10$ or $2/5$

Replace the value of x in original problem by the value of x.

$3x - 1/5 = x + 3/5$

$3(2/5) - 1/5 = 2/5 + 3/5$

$6/5 - 1/5 = 2/5 + 3/5$

$5/5 = 5/5$ or $1 = 1$ (true equation)

Solve the same problem by adding and/or subtracting items from both members before you multiply/divide. Use your discoveries to decide which method is a more appropriate method for solving equations.

Solve example 1 using addition and/or subtraction.

$3x - 1/5 = x + 3/5$

$3x - 1/5 + 1/5 = x + 3/5 + 1/5$ (1/5 was added to both members.)

$3x = x + 4/5$ (Simplified.)

$3x - x = x + 4/5 - x$ (Subtract x from both members.)

$2x = 4/5$ (Simplified.)

$1/2(x) = 1/2(4/5)$ (Multiply both members by 1/2, the reciprocal of 2.)
$x = 4/10$ or $2/5$

Remember that the only quantities that can be combined are those with a common name. Notice that the solution of the equation is an application of the properties of numbers discussed earlier. They can be applied in any order as long as they are not violated. You may wish to solve the examples in other ways without violating any of the basic properties of the number system studied earlier.

Remember that the division bar (—) is written as / on the computer and will be used henceforth.

Example 2: Solve: $9x + 10 + 5x + 14 = 12x - 18$. Begin by combining all similar items in each member.

$9x + 10 + 5x + 14 = 12x - 18$
$(9x + 5x) + (10 + 14) + 18 = 12x - 18 + 18$ (Use grouping and equation properties.)
$14x + 42 = 12x$ (Simplify by combining terms in each member)
$(14x - 12x) + 42 - 42 = (12x - 12x) - 42$ (Use equation properties and grouping.)
$2x = -42$ (Simplify by combining terms in each member)
$1/2(2x) = 1/2(-42)$ (Use reciprocals to find the value of x.)
$x = -21$ (factoring and using cancellation or by use of reciprocals.)

Check to see if equation is true by substituting in original equation.

Example 3: Solve: $2x + 5 = 11$ by using methods 1 and 2 suggested below.

1. Begin by adding –5 to both members.
2. Begin by multiplying by the reciprocal of 2.

Compare the solutions and make a prediction for the better process to use when solving equations.

Use your preferred solution option to solve the following equations.

Example 4: Solve $9x + 4 = 6$.
Example 5: Solve $6x + 15 = 8x - 21$.

Writing Equations

Most problems encountered will be in your native language and must be translated into the language of mathematics for resolution. A method is listed below that may be used in the translation process. You may develop your own best method of approach, but it is suggested that you seek to understand the process with some format that can be modified at your discretion without violation of any of the principles of mathematics.

Suppose Jim is twice (2 times) as old as Mary, and their combined age is seventy-five years. How old is Jim, and how old is Mary?

To solve this problem, you must ascertain the name of the final quantities that are to be sought after and to identify the information provided in the statement of the problem.

1. Do you know the age of either Mary or Jim?
2. Can their ages be represented by some quantity either known or unknown?
3. If you don't know their ages, then what do you know about their ages?
4. Can you create a visual model?

For question # 1, their actual ages are not known.
For question #2, yes, if Mary's age is represented by x, then Jim's age is 2x. Conversely, if Jim's age is represented by x, then Mary's age is x/2 or 1x/2.
For question #3, you know the sum of their ages is seventy-five years, and the sum of their ages is represented by Mary's age added to Jim's age.
For question #4, yes, you can create a statement of the sum of their ages. Therefore, Mary's age added to Jim's age is seventy-five years; $x + 2x = 75$ is the sought-after equation that represents the information in the original statement when Mary's and Jim's ages are represented as follows:

x = Mary's age
$2x$ = Jim's age
$x + 2x$ = sum of their ages
75 = sum of their ages
$x + 2x = 75$, or if
x = Jims' age, then
$x/2$ = Mary's age
$x + x/2$ = sum of their ages
75 = sum of their ages
$x + x/2 = 75$

Example 2: The perimeter of a rectangle, whose length is three times its width, is 150 meters.
Determine the length and width of the rectangle.

1. Do you know the actual length or width of the rectangle?
2. Can their dimensions be represented by a known or unknown quantity?
3. If you don't know the dimensions, is there a relationship between them?
4. Can you create a model?

For question #1, no, the actual length and width are not known.
For question #2, yes, the length is three times the width.
For question #3, yes, the perimeter is 144 meters (perimeter is the distance around the surface and is the sum of the lengths and widths of the rectangle)
For Question #4, yes, you can draw and label a rectangle to visualize the conditions set forth.

The distance around the figure is the perimeter, and there are two lengths and two widths to be considered when referring to the perimeter.

Solution:

If w = the width, then 3w = the length.
Perimeter = 2 widths + 2 lengths (definition of perimeter). *Peri* means *around*, and *meter* means *to measure*.
2w +2(3w) = the perimeter of the rectangle.
144 = the perimeter of the rectangle.
2w +2(3w) = 144; continue with the solution.

Notice that the solution can be found when mathematical statements written as their language equivalent allow for the writing of the proper equation from which a solution can be found.

Procedure for Solving Problems

1. Read the statements to ascertain what the problem is asking (identify and assign names).
2. Identify the designated items and assign values to them (quantify and assign values).

3. Compare statements to find statements that are exactly alike (analysis: compare the statements).

All statements should be clearly determined and written in plain English. That becomes an equation to be solved.

Example 1: Two boys went fishing, and one boy caught one less than twice (two times) as many fish as the other boy. If they caught a total of twenty-two fish, how many fish did each boy catch?

A. Identify and assign designated names
    a.   ? = the number of fish caught by the first boy
    b.   ? = the number of fish caught by the second boy
    c.   22 = the total number of fish caught by both boys
B. Quantify, assign values to each item in #1 above
    a.   If x = the number of fish caught by the first boy,
    b.   then $(2x - 1)$ = number of fish caught by the second boy.
    c. $x + (2x - 1)$ = the total number of fish caught by both boys.
    d.   22 = the total number of fish caught by both boys.
C. Analyze and compare the statements from above to find comparable statements and write an equation.
    a. $x + (2x - 1)$ = total number of fish caught by both boys
    b.   22 = total number of fish caught by both boys

The equation becomes $x + (2x - 1) = 22$.

Example 2: Two boys wanted to pool their money to buy a video game that cost $235. One boy had five dollars less than twice the amount of the other boy. How much did each boy contribute to the purchase of the video game?

A. Identify and assign names
    a.   ? = amount contributed by the first boy
    b.   ? = amount contributed by the second boy
    c.   235 = total amount contributed by both boys
B. Quantify—assign values
    a.   If x = amount contributed by the first boy,
    b.   then $(2x - 5)$ = amount contributed by the second boy.
    c. $x + (2x - 5)$ = total amount contributed by both boys.*
    d.   235 = total amount contributed by both boys.*
C. Analyze—find statement that describe the same items
    a. $x + (2x - 5)$ = total amount contributed by both boys.*

      b.   235 = total amount contributed by both boys.*
           235 = x + (2x – 5).

Why is 2x – 5 written as (2x – 5)? Review the symbols of inclusion.

Dimensions and Trig Functions on Plane Surfaces

If several basic number lines were laid, one on top of the other,

1. How many would you see?
2. Would you be able to tell how many lines are laid together?
3. If your answer is one (1), why is that so?
4. Do the lines exist—or is it that you cannot see them or detect their existence?
5. Now, suppose there were only two lines present, and one of the lines began to rotate from right to left. At that time would both lines become visible?
6. If the number line represents one dimension, would the second line also represent one dimension? If so, then why? If not, then why not?

For question #1, you would see only the x-axis. The separate line(s) would not be visible as a separate dimension. It would become apparent only after it changed positions.
For question #2, there are two possibilities. One possibility is that the line on top obscured the lines underneath, and the second possibility is that the second line(s) represents a quantity too small to be visible. (Think about that possibility.)
For question #4, both possibilities exist, and both are plausible.
When working on surfaces, a new term becomes necessary. *Dimension* is a term used to designate a line that moves in only one direction from a given reference point and is referred to as a reference line when it originates at the point called "0."

Dimension is a line that exists (moves) in only one direction from a reference point. A number line is a one one-dimensional entity (a horizontal line with no vertical movement). The number line listed below can be represented as distances on the line as illustrated below:

0, 1, 2, 3, 4, 5, 6, 7, 8, 9, 10, 11, 12, 13, 14, 15, 16, 17, 18, 19, 20, 21, 22, 23, 24, 25, 26, 27, 28, 29, 30, 31, 32, 33, 34

Note the uniform distances represented by each numeral below:

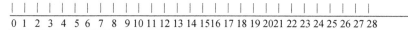

Notice that points on that line can be named with one numeral. If an identical line was to be rotated counterclockwise (right to left) where the x numeral is always 0, it would also move in only one direction and would represent a second reference dimension with no horizontal movement.

Examination of the diagram below will show that as the OP moves counterclockwise, the point P moves closer and closer to the reference point 0. When the line rises directly above 0, you will notice that another line will have been created that moves in only one direction and is also a reference dimension. You will now see that we have two reference dimension lines: one horizontal (flat) and one vertical (straight up on the page).

Study the illustration below to see the reference lines OX and OY that move in only one direction from a common reference point. Each reference line represents a *dimension* to which all movement on a surface can be referenced. Motion in all directions can be determined by using numerals on the dimension lines.

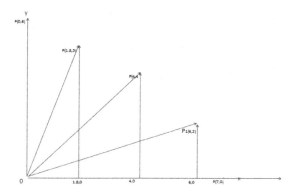

The lines 0X and 0Y represented above are dimensions that will be used as reference dimensions. The lines are given names of x and y and are called axes. The line named $0P_1$ is moving in two directions relative to the axes x and y and is moving in two dimensions. Notice that as line $0P_1$ rotates upward, the point "p" moves both horizontally and vertically and is two-dimensional. Can you find another dimension? (Create a model to illustrate this idea.) Notice that if $0P_1$ were placed directly on top of the x-axis, it would not be visible as a separate dimension. It would become apparent only after it changed positions. Did it exist even if it could not be seen (detected)—or was it invisible because it was too small to be seen (detected)? Using your answer, how many dimensions can exist (2, 3, 4, 5 ... unlimited)?

Dimensional lines exist, and have the same value, as each number on the basic number line, which means there are unlimited dimensional lines with each arising from a reference number on the basic number lines. Since reference lines x and y move in only one direction relative to a reference point "0," any other line that crosses both vertical and horizontal dimensional lines must travel through more than one dimension. Straight lines that cross dimensional lines are two dimensional. Two dimensional lines are named using both reference dimensional lines when compared to the reference lines (axes) x and y and must be named using a modified naming system using numeral names from each reference axis (x and y as shown below). The names of the points in a two-dimensional statement are called *coordinates* and are listed in the order of x and y, with x always listed first. The proper listing of the coordinates is always (x, y). The points listed as (x, y) are called coordinates and must always be listed in the (x,y) format.

In the new naming system, all points are named with two numerals instead of one as before. This system allows us to compute in a two-dimensional space instead of only one as before. A two-dimensional space is called a *plane surface* and is physically represented by a flat surface. Note also that a plane surface must contain at least two lines and must contain items from both axes (the x and y dimensions).

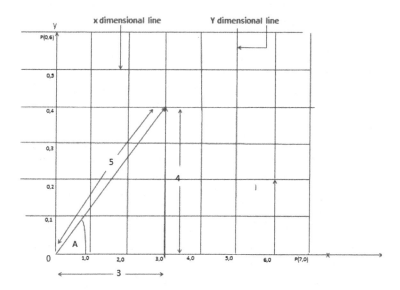

Circles and Trig Functions

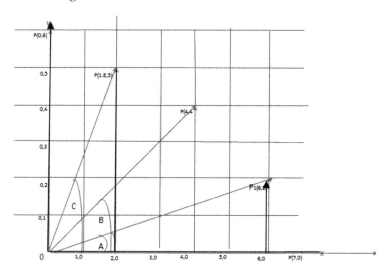

In the diagram above, as point p moves upward to positions p_1, p_2, p_3, the point p moves in a circular pattern. The circular distances A, B, and C measured from the base number line are called *angles* and are measured in circular measures called *degrees*. The circular motion of degrees can be related to the basic dimensions using the basic number lines that represent the dimensions of horizontal distance and vertical distance.

The basic structure for measurement of angles is the circle, which is divided into 360 equal basic parts (degrees). Each degree is divided into 60 equal parts (minutes), and each minute is divided into 60 equal parts (seconds).

Some new terms to be used in this relationship are:

1. Angle: the circular measure of two lines represented in two dimensions (expressed in a unit called degrees)
2. Circle: the path of a curved line that is always the same distance from a given point ◯
3. Radius: the distance from the center to the curved line of the circle
4. Diameter: line passing through the center and intersecting the circle in two distinct points
5. Degree: one part of a circle, which consists of 360 equal parts
6. Minute: one part of a degree, which consists of 60 equal parts
7. Second: one part of a minute, which consists of 60 equal parts
8. Vertical distance: the distance from a point on the "x" dimension line to a point on a "y" dimensional line
9. Horizontal distance: the distance from a point on the "y" dimension line to a point on the "x" dimensional line
10. Sine of an angle (sin): the ratio (fraction) formed by the vertical distance to a point divided by the radius of the circle that contains that point
    - Sin A = vertical distance to a point
    - radius of the circle that contains that point
11. Cosine of an angle (cos): the ratio (fraction) formed by the horizontal distance to a point divided by the radius of the circle that contains that point
    - Cos A = horizontal distance to a point
    - radius of the circle that contains that point
12. Tangent of an angle (tan) = the ratio formed by the vertical distance to a point divided by the horizontal distance of the circle that contains that point
    - Tan A = vertical distance to a point
      horizontal distance to a point
13. Slope: same as the tangent

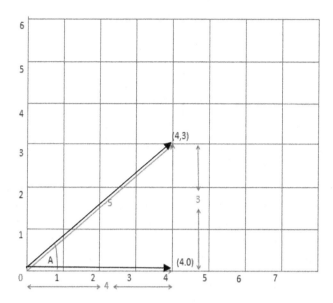

Sin A is the name of the ratio 3/5; sin A = y/r.
Cos A is the name of the ratio 4/5; cos A = x/r.
Tan A is the name of the ratio 3/4; tan A = y/x.

Slope is another name for the tangent, which is 3/4 in the example above. Slope is often expressed as rise over run or *m = rise/run* where m is the name given to the slope of a line. The slope of the line above implies that for every four units traveled on the x-axis, three units will be traveled on the y-axis—and the slant of the line can be accurately measured.

Since Sin A is a number represented as a ratio, it has a reciprocal such that their product equals 1. The reciprocal of sin A is called secant (sec) A and (sin A)(sec A) = 1. If sin A is denoted by symbols y/r, can you determine sec A? Cos A = x/r, cosecant (csc) A is the reciprocal of cos A and (Cos A)(csc A) = 1.

The equation x + y = 10 is a statement that exists in two dimensions, and the sum of the coordinates equals 10. If the sums are plotted on a graph as shown below, they will form a

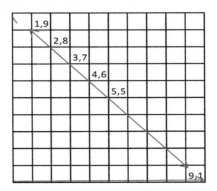

unique line. Since every line can be paired with numbers on dimensional lines (number lines) and the numbers are infinite and unique, the number of lines is also unlimited and unique. Every line has a unique name written in the form $ax + by = c$ where a, b, and c are known quantities. For example, $2x + 3y = 12$ is the name of a unique line, which means there is no other line with that exact name.

Since lines can be represented as numbers, they can be used in operations just the same as numbers. If $2x + 3y = 12$ is the equation of a unique line, then the equation can be manipulated. If no tenant of the system is violated, the equation can be transformed by using equation principles.

If $2x + 3y = 12$, then $4(2x + 3y) = 4(12)$ is still a valid equation.

If $2x + 3y = 12$, then $2x + 3y - 3y = 12 - 3y$ is still a valid equation.

If $2x + 3y = 12$, then $\dfrac{2x + 3y}{4} = \dfrac{12}{4}$ is still a valid equation.

Example: If $ax + by = c$,

$by = c - ax$ is still a valid equation or

$by = -ax + c$ is still a valid equation

$b(1/b)y = (-ax + c)(1/b)$ is still a valid equation

$y = \dfrac{-ax + c}{b}$ is still a valid equation.

That is an example of ways the structure of the system of mathematics influences all other uses of the system. That is one of the strengths of the system, and it makes it very a reliable, non-controversial system that is used to make reliable predictions in other closed systems.

Conclusion

This process could continue almost indefinitely to encompass many different phases of math, but that is not the purpose of this document. This book is intended to acquaint the reader with the basic structure of the language of math and to show how processes and procedures developed are connected to the central idea of the structure. Another aim is to show implications of the structure to a basic understanding of how math is constructed and to show that the structure of math is basically similar to the structure of any language (with a few notable exceptions). A basic understanding of the concepts will demystify the fundamentals of math and eliminate the need for many of the processes that are developed and used to solve problems.

This book does not attempt to solve problems. The intent is to present basic fundamentals of the system and their relationships to the structure so that readers can develop processes and procedures that rely on an understanding of the system rather than memorizations of basic facts, such as "a negative times a negative equals a positive" or questions such as "what are the steps?" A basic understanding of the structure will diminish the need for acronyms such as "Please Excuse My Dear Aunt Sally" or "FOIL" in working with numbers. A basic understanding of the structure and concepts outlined in the system will make transferring to other mathematics-based systems much easier and less confusing. When working in sciences such as chemistry or physics, a basic understanding of the power of zero and one will help make the work much easier to accomplish.

Some examples could be included, but that is not the premise of this book. I hope readers will take the time and effort to fully comprehend the concepts put forth and make some effort to create situations where they can be applied as an aid to a fuller comprehension of the contents.

A second section has been added with some exercises and ideas that are intended to reinforce the concepts demonstrated thus far. Please refer to the following pages that are included with this effort.

Comprehensive Review of All Mathematical Concepts

Addition: counting the numerical part of numbers (movement on the number line)

Addition Symbol(s) + (and): to continue moving in the same direction as the original direction

Additive Identity: addition of "0" to a number can change its appearance but not its value

Absolute Value: the positive or negative distance that a number is from the reference point (0)

Associative Property: when adding numbers, they may be grouped in any convenient manner with no change in value

Basic Number Line: the number line from which the "answer" comes: 2(2) = 4(1) or 1(4)

Basic Symbols: 0, 1, 2, 3, 4, 5, 6, 7, 8, 9

Base: the number that is raised to a power

Binary: consists of two parts (all basic numbers are binary)

Basic Number: consists of two parts with a numerical part and 1: 1(0), 1(2), 1(9) (written as groups)

Cancellation: division of numbers with the same value or multiplication by the reciprocal

Closure Principle: all items and conclusions must be contained within the system of mathematics or any closed system

Closure Principle: All operations performed on members of the system will produce a result that is also a member of the system. Everything that happens must be contained within the system. It also allows for the making of predictions of events within the system and serves as a model for making predictions in other closed systems.

Co-efficient: The counting part of a number; The naming part is the base; 2x

Composite Number: the result of counting (adding) numbers

Counting: moving along the numerical sequence

Decimal: the result obtained when a fraction is written in another form by continued division

Dimensions: lines that move in only one direction from a reference point

Dimensional Lines: lines that originate at each number on the basic number line(s)

Distributive Property: addition of numbers to form composite numbers
Example: 3(2) +5(2) = (3 + 5)(2) and by symmetric property (3 + 5)(2) = 3(2) +5(2)

Distributive Property: counting numbers to produce composites
Example: 2(3) + 5(3) = (2 + 5)(3) or by symmetric property (2 + 5)(3) = 2(3) + 5(3)

Division: the number of times one number can be subtracted from another number

Division of Powers: to divide with powers, write the base and subtract exponents

Exponent: the raised number that represents the number of factors contained within a number

Factor: the parts of a number of composite number

Integers: the new group of numbers that include the whole numbers and the negative numbers

Multiplication: writing of numbers or composite numbers

Multiplication Identity: Multiplication by one can change the appearance of a number but does not change its value or size

Multiplication of Powers: to multiply with powers, write the base and add exponents

Negative Numbers: numbers to the left of zero on the number line (predicted by the closure principle)

Number Line: the arrangement of the numerals into a sequence increasing or decreasing by unit values

Numeral: the names assigned to the basic symbols

Power: the number of times a factor appears in a number

Prime Number: a number that has one set of factors (one of which is always 1)

Product: the result of multiplication of numbers (the answer from the basic number line)

Reciprocal: the result obtained when the factors of a number produce a product of "1"

Reference Point: a point of origin that is designated as "0" on the number line

Reflexive Property: the parts of a number can be interchanged without changing the value

Radius: the distance from the center of a circle to the circle (curved line)

Root: numbers that are raised to a fractional power:

> Square Root: Numbers that are raised to a power of 1/2
> Cube Root: Numbers that are raised to a power of 1/3
> Fourth root: Numbers that are raised to a power of 1/4

The powers can be any numbers including fractions.

This can be shown by writing the individual factors and using cancellation. How?

Subtraction Symbol (–) (difference): means to move in the opposite direction from the original direction

Sum: the process of addition to find a total amount

Symmetric Property: property where quantities have different appearances but have the same value (the basis for the concept of equations)

Bobby Rabon

Example: $1(1) + 1(1) = (1 + 1)(1) = 2(1)$: Notice that $1(1) + 1(1)$ and $2(1)$ represent the same quantity but do not have the same form.

Transitive Property: When each of two quantities is equal to a third quantity, the two quantities are equal to each other.

Example: If $3(4) = 12$ and $1(2) + 5(2) = 12$, then $3(4) = 1(2) = 5(2)$.

Term: another name for a number in an equation

Whole numbers: numbers that begin with 0 (move unit distances to the right)

Writing of Numbers: numbers are written with grouping symbols

Lagniappe

(A Little Something Extra)

Meaning of Numbers

Numerals are names of items on the number line, and numbers are the results of counting the objects on the number line. Numbers always consist of two parts: a counting part and a naming part.

> Example: 2(3); 2 is the counting part, and 3 is the naming part.
> Example: 2(ants); 2 is the counting part, and ants is the naming part.
> Example: 3(ants) + 4(ants) = (3 + 4) ants; 3 and 4 are the counting parts, and ants is the naming part.
> Example 5: x(squares) + y(squares) = (x + y)squares; x and y are the counting parts.

The number line is created by repeating the basic numeral alphabet and following the pattern of grouping into named groups (each contains all of the basic alphabets).

The basic numerals are 0, 1, 2, 3, 4, 5, 6, 7, and 9; the number assigned to the basic symbols is the same as the number named ten (10). The basic symbols are used to construct the basic system by repeating the basic numeral in a sequence that creates a pattern with which all basic numbers are created.

Group 0	Group 1	Group 2
|0,1,2,3,4,5,6,7,8,9|10,11,12,13,14,15,16,17,18,19|20,21,22,23,24,25,26,27,28,29|..................
Group 3	Group 4	
|30,31,32,33,34,35,36,37,38,39|40,41,42,43,44,45,46,47,48,49|...	

Group 5 ....Group 9
|50,51,52,53,55,56,57,58,59|... 90,91,92,93,94,95,96,97,98,99|
Group 10
|100,101,102,103,104,105,106,107,108,109|
Group 50   Group 100
.... 500,501,502,503,504,505,506,507,508,509__|1000,1001,1002,1003,........1009|

Note that that the number line is created continually by adding basic groups of numerals and following the basic established pattern.

Counting is moving along a numerical sequence, and adding is moving along a number sequence.
Counting numbers produces new numbers called composite numbers.

Addition (tentative definition) is moving from left to right on the number sequence, and subtraction is moving from right to left on the basic number sequence.

Addition/Subtraction

a. $2(1) + 3(1) = (2 + 3)(1)$ or $(3 + 2)(1)$
b. $3(2) + 4(2) = (3 + 4)(2)$ or $2(3 + 4)$
c. $a(c) + a(c) = 1(ac) + 1(ac) = (1 + 1)(ac) = 2ac$
d. $a(c) + b(c) = (a + b)c$ or $c(a + b)$
e. $1(a + b)(c + d) + 2(a + b)(c + d) = (1 + 2)(a + b)(c + d) = 3(a + b)(c + d)$
f. $3(2) - 2(2) = (3 - 2)(2) = 1(2)$ or $2(1)$
g. $2(2) - 3(2) = (2 - 3)(2) = -1(2) = -2(1)$ or $2(-1)$: illustrates that subtraction is not commutative

Addition is counting numbers to produce new numbers called composites. Subtraction is the addition of the opposite of the number after the subtraction sign. Multiplication is the writing of numbers or composites both of which always consist of two or more factors.

a. $3(1)$ factors are 3 and 1 prime (one set of factors)
b. $8(1)$ factors are 8 and 1 or 4 and 2 composite (multiple sets of factors)
c. $a(b + c)$ factors are a and $(b + c)$ prime (one set of factors)
d. $ab(c + d)$ factors are a, b, $(c + d)$ composite (multiple factors)
e. $3(3)$ factors are 3 and $3 = 3^2$ power (same factors)
f. $5(5)(5)$ factors are 5, 5 and $5 = 5^3$ power (same factors)
g. $(a + b)(a + b) = (a + b)^2$ factors are $(a + b)$ and $(a + b)$

Division is the number of times one number can be subtracted from another number with the same name.

a. a. $8/2 = 4$, which is the number of 2(1) can be subtracted from 8(1)
b. b. $30/5 = 6$, which is the number of times 5(1) can be subtracted from 30(1)
c. c. $(a + b)^2 / (a + b)^1 = (a + b) / (a + b)$ (2) times

There is also a division algorithm that will not be demonstrated but can also be used. The basic properties of the system must be internalized and understood to become competent problem solvers in this system of mathematics. Division can also be written as multiplication by the reciprocal of the denominator.

$$\frac{10}{5} = \frac{10(1)}{(5)} \qquad \frac{5}{6} = \frac{5(b)}{6(a)}$$
$$\frac{a}{b}$$

Reflexive Property

The order in which a number is written does not alter its value or meaning.

a. $1(2) = 2(1)$, $xy = yx$, $(a + b) = (b + a)$

Symmetric Property

In this form of the reflexive property, each side has a different appearance. One side can be a basic number, and the other can be a composite with the same value.

a. If $3(1) = (2 +1)(1)$, then $(2 +1)(1) = 3(1)$.
b. If $x = (y + z)$, then $(y + z) = x$.
c. If $a = b$, then $b = a$.

Commutative Property of Addition

When adding numbers, the order in which they are added does not alter the outcome. Since multiplication can be a form of addition, it is also commutative.

a. $2 + 3 = 3 + 2$
b. $a + b = b + a$

c. $xyz + pqr = pqr + xyz$ (parenthesis is omitted when dealing with such named numbers)
d. $2(x - 3y) + 3(4x + 5y) = 3(4x + 5y) + 2(x - 3y)$

Commutative Property of Multiplication

Multiplication is a form of addition and is also commutative.

a. $2(3)(4) = 3(2(4)$ or $4(3)(2)$
b. $x(a + b)c = (a + b)xc$ or $xc(a + b)$

Associative Property of Addition

When adding numbers, the manner in which they are grouped does not alter the outcome. It is also true for multiplication.

a. $(2 + 3) + 4 = 2 + (3 + 4)$ (Review the meaning of symbols of inclusion.)
b. $p + (q + r) = (p + q) + r$
c. $[a + (b + c) + (x + y)] + pq = [a + (b + c) + pq] + (x + y)$

Associative Property of Multiplication

Multiplication is a form of addition and is also associative.

a. When adding numbers, the order in which they are computed does not alter the outcome. (Remember that multiplication and addition are binary operations.)
b. When multiplying numbers, the order in which they are computed does not alter the outcome.

Division is a form of subtraction (deletions) and the order of the operations cannot be change
$(3 - 4) = -1$ and is not the same as $(4 - 3) = 1$

Distributive Property

Counting numbers to produce a composite (see addition of numbers).

  a. a. $2(3) + 4(3) = (2 + 4)(3) = 6(3)$ and by symmetric property $6(3) = (2 + 4)(3)$
  b. $2(3) - 4(3) = (2 - 4)(3)$ or $3(2 - 4)$
  c. $ab(c) + pq(c) = (ab + pq)(c)$ (Review the symmetric property.)

Transitive Property

This allows for substitution by identifying quantities of equal value.

  a. If $3 = (2 + 1)$ and $(2 + 1) = (7 - 4)$, then $3 = (7 - 4)$.
  b. If $2x = 3p$ and $3p = 3x - 15$, then $2x = 3x - 15$.
  c. If $(p + q) = 5xy$ and $5xy = (10p - 4q)$, then $(p + q) = (10p - 4q)$.
  d. This property allows for substitution of items, which can allow for operations with items that are dissimilar in appearance.

Reciprocals

The value of the product of numbers is always positive 1.

  a. If 2 is a numeral, then $2(\frac{1}{2})$ is a number whose product is always 1.

Note: $\frac{2(1)}{(2)} = \frac{2}{2} = 1$ is also a form of division where 2 can be subtracted 2 exactly 1 time.

Note also that the reciprocal appears to invert (flip upside down) when solved.

Therefore, $\frac{3/4}{3/4} = \frac{3(4)}{4(3)} = \frac{12}{12} = 1$.

Since the outcome is always +1, you must make use of the rules for working with positive and negative numbers

  b. Reciprocals must always be either both positive or both negative.
      $\frac{-3(4)}{4(-3)} = 1$

Multiplicative Identity

Multiplication by 1 does not change the value of number and can be used to change the name or appearance of a number by the use of reciprocals.

a.   If you know that 2 cups = 1 pint; 1 quart = 2 pints, and 1 gallon = 4 quarts, you can use multiplicative identities to find the number of cups in 4 gallons.

$\dfrac{(2 \text{ cups})}{(1 \text{ pint})} = 1$ and $\dfrac{2 \text{ pints}}{1 \text{ quart}} = 1$ and $\dfrac{1 \text{ gallon}}{4 \text{ quarts}} = 1$

4 gallons = 4 gallons$\dfrac{(4 \text{ quarts})(2 \text{ pints})(2 \text{ cups})}{(1 \text{ gallon})(1 \text{ quart})(1 \text{ pint})} = \dfrac{4(4)(2)(2)}{1}$ cups = 64 cups

Note that from the known information that 2 cups = 1 pint; 2 pints = 1 quart;

4 quarts (qts) = 1 gallon (gal) can be shown equal to 1 by solving the equation
4 qts = 1gal

$\dfrac{4 \text{ qts}(1)}{1 \text{ gal}} = 1 \text{ gal}\left(\dfrac{1}{1 \text{ gal}}\right)$ and it is shown that $\dfrac{4\text{qts}}{1 \text{ gal}} = 1$

The computed values can be used in the same manner as "1."

b. If you know that 1 inch = 2.54 cm; 1 cm = 10 mm; 1 ft = 12 in, then the question of how many mm are there in 500 ft can be solved.

500 ft = 500 ft $\dfrac{(12 \text{ in})(2.54 \text{ cm})(10 \text{ mm})}{(1\text{ft}) (1 \text{ in}) (1 \text{ cm})} = \dfrac{500(12)(2.54)(10)}{1(1)(1)} = 152400$ mm

Inspection of the equations above will show that the conversions were accomplished by repeated use of factors with a value of 1.

Additive Identity

Addition of "0" does not change the value of a number, but it can change the appearance of the quantity.

  a. $8 + 0 = 8 + (5 - 5) = (8 + 5) - 5 = 13 - 5$ (Appearance changed to $13 - 5$)
  b. $a + 0 = a + (b - b) = (a + b) - b$ (Appearance changed to $(a + b) - b$, but the value is still the same, which is originally called simply "a".)

Closure Principle

Everything that happens within the system produces a result that lies within that system. An example is the system of mathematics and, as far as is known, the universe. Closed systems have a commonality that allows one, the mathematical system, to serve as a model for the other, the universe, and other systems that are closed.

  a. If 5 is on the number line and 3 is on the number line, the sum 5 + 3 is also on the number line.
  b. Any operation with numbers will produce a result that is also a number that must be unique.

Symbols of Inclusion

Everything that is enclosed within the symbols of inclusion—no matter if it consists of one part or many parts—is considered to represent one number or term.

  a. $\{ \}$, $[\,]$ and $(\,)$, are symbols of inclusion and can be bundled into multiples as shown below.
     1. $\{2 + [3 + x(x + 3)] - 5\}$ represents one number (remember + and – signs).
     2. $2(3)$ represents one number or term.
     3. $\{2(3) + (x - y) - 6\}$ represents one number or term.
  b. Other symbols of inclusion are
     1. division bar —
     2. absolute value symbol $|\ \ |$

Reference Point (0)

This separates the number line into equal parts and determines the names of the numerals in each part. One noteworthy idea is that the sum of all the numbers in the number line is equal to zero (0). What do you think of that statement?

Integers consist of positive and negative numbers and do not contain fractions.

Absolute value is the distance a numeral is located from the reference point (0) and always has a positive value. It is denoted by the symbol $|\ |$. Therefore:

a. $|2| = 2$ since 2 is a positive number.
b. $|-2| = -(-2) = 2$ since (-2) is a negative number.

The conditions expressed are generally written and must always be considered.

$|a| = a$ if a represent a positive number.
$|a| = -a$ if a represents of a negative number.
$|a| = 0$ if $a = 0$.

Example 1: If $q = (x + y)$, then $|q| = |(x + y)| = (x + y)$ if $(x + y) > 0$.
If $q = (x + y)$, then $|q| = |(x + y)| = -(x + y)$ if $(x + y) < 0$.
If $q = (x + y)$, then $|q| = 0$, if $(x + y) = 0$.

The Meaning of the Symbols "+" and "−" Signs

Subtraction has been defined as defined as "addition of the opposite" of the number after the subtraction sign. This implies that the operation of subtraction is to be treated as addition, which implies that all counting begins at the origin. Therefore, a statement such as $5 + 6$ can actually be written as $0 + 5 + 6$, and the statement $- 5 + (-6)$ can actually be written as $0 + (-5) + (-6)$ and all negative numbers must be enclosed in a symbol of inclusion. The statement $-5 + 6$ implies that 5 is a positive number and the − sign means that the direction must be reversed. All negative numbers are to be enclosed in symbols of inclusion.

a. A "+" sign means that the direction of the number is unchanged when the operations of addition and multiplication are performed.

1. If 2 is moving in the positive direction when compared to the reference point "0," then $2 + 2 = 0 + 2 + 2 = 4$ (the direction of the 2 after the + sign did not change.)

2. If 2 is moving in the negative direction when compared to the reference point "0," then $2 + (-2) = 0 + 2 + (-2) = 0$ (the direction of the 2 after the + sign did not change).

3. If 2 is moving in the negative direction when compared to the reference point "0," then $(-2) + (-2) = 0 + (-2) + (-2) = -4$ (the direction of the 2 after the + sign did not change).

4. If 7 is moving in the negative direction when compared to the reference point "0," then $-7 + (-3) = 0 + (-7) + (-3) = -10$ (the direction the numbers after the + sign did not change).

b. A "–" sign means that the direction of the number is reversed when the operations of addition and/or multiplication are performed.

1. If 2 is moving in the positive direction when compared to the reference point "0," then $2 - 2 = 0 + 2 + (-2) = 0$ (the direction of the 2 after the – sign did reverse directions). Remember that both of the 2s are positive unless written as (-2).

2. If 2 is moving in the negative direction when compared to the reference point "0,"
then $2 - (-2) = 0 + 2 + 2 = 4$ (the direction of the 2 after the – sign did reverse directions).

3. If 7 is moving in the positive direction when compared to the reference point "0," then $3 - 7 = 0 + 3 + (-7) = -4$ (the direction of the 7 after the – sign did reverse directions).

4. If 7 is moving in the negative direction when compared to the reference point "0," then $-3 - (-7) = 0 + (-3) + 7 = 4$ (the direction of the 7 after the – sign did reverse directions).

Bobby Rabon

Multiplication of Positive and Negative Numbers
When comparing the multiplication of positive numbers with the multiplication of negative numbers, it can be discerned that the same process patterns should be the same for both types as illustrated in the examples below.

    a. Since the multiplication of two positive numbers results in a positive outcome, then by utilizing the same pattern, it can be discerned that the multiplication of two negative numbers will also produce a positive outcome by application of the meaning of the negative sign before a number.

1. $3(2) = 0 + 3(2) = 6$ (The + sign before a number means that the direction is not reversed.)
2. $3(-2) = 0 + 3(-2) = -6$ (The + sign before the number did not reverse the direction of the number.)
3. $-3(-2) = 0 - 3(-2) = 0 + 3(2) = 6$ (The negative sign before the number reversed the direction of the number—note that the first 3 is positive, and the negative sign preceding it reversed the direction of the number.)
4. $-3(-2)(-4) = -24$ (multiplication is binary and $6(-4) = -24$).

Multiplication with the Definition of Multiplication by Addition

a. Multiply 16 by 4 or 16(4).
$$16 = 10 + 6 \text{ (by application of the distributive property)}$$
$$4(10 + 6) = 4(10) + 4(6)$$
$$= 40 + 24$$
$$= 64$$

b. Multiply 98(102)
$$98 = (100 - 2) \text{ and } 102 = (100 + 2)$$
$$98(102) = (100 - 2)(100 + 2)$$
$$= 100(100) + 100(2) - 2(100) - 2(2)$$
$$= 10000 + 200 - 200 - 4$$
$$= 9996$$
Or it could have been done as
$$98(102) = 100(100 + 2) - 2(100 + 2)$$
$$= 10000 + 200 - 200 - 4$$
$$= 9996$$

c. Multiply 635(15)

635 = (600 + 30 + 5), 15 = (10 + 5)

635(15) = (600 + 30 + 5)(10 + 5)

$$= 600(10) + 600(5) + 30(10) + 30(5) + 5(10) + 5(5)$$
$$= 6000 + 3000 + 300 + 150 + 50 + 25$$
$$= 9525$$

Or it could have been done in many other ways by application of math principles.

d. Multiply 635(15) again.

635 = (600 + 30 + 5) and 15 = (10 + 5)

635(15) = 10 (600 + 30 + 5) + 5(600 + 30 + 5)

$$= 6000 + 300 + 50 + 3000 + 150 + 25$$
$$= 9525$$

These examples illustrate how the use of the basics of the system can be used in various ways if no parts of the system are violated, and the solutions to problems depend upon the insight of the responsible person. With an understanding of the system, many operations can be performed using your insights as well as the algorithms outlined in textbooks and other venues.

e. Multiply 8(9).

8(9) = (10 − 2)(10 − 1) because 8 = 10 − 2 and 9 = 10 − 1.

8(9) = (10 − 2)(10 − 1)

$$= 10(10 − 1) − 2(10 − 1)$$
$$= 100 − 10 − 20 + 2$$
$$= 72$$

Examination of the examples above will show that multiplication can be performed by basically using 2, 5, and 10 (or multiples of 10). Think about that and try other examples.

Division of Positive and Negative Numbers

1. Since multiplication is the writing of numbers or the combining (adding) of numbers to form composites and division is subtracting of number to find their basic components, the rules for multiplication and division of numbers obey the same rules as above.

Divide 32 by 6 (note that 32 can be written as 12 + 12 + 6 + 2).

$$\frac{2 + \ \ 2 + \ \ 1 + 2/6}{6 \overline{)12 + 12 + 6 + 2}} = 2 + 2 + 1 + 2/6 = 5 + 1/3$$

Divide 750 by 25.

$$\frac{4\ +4\ +\ \ 4\ +\ 4\ \ +4\ +\ \ 4\ \ +\ \ 4\ +\ 2}{25\lfloor 100\ +\ 100\ +\ 100\ +\ 100\ +\ 100\ +\ 100\ +\ 100\ +\ 50} = 30$$

Divide 9872 by 327.

$$\frac{10\ +\ \ 10\ +\ 10}{327\lfloor 3270 + 3270 + 3270 + 162} = 30 + \frac{162}{327}$$

2. There is a division algorithm, but there are numerous ways to divide numbers. Some will be illustrated below, but I will not list the division algorithm because I want you to understand that division is not limited to the algorithm. It can be accomplished in other ways and by other means if you fully understand the meaning that has been ascribed to the process we call division.

    a. Let us divide 6762 by 6.
        1. Since 6762 has four numerals, we can multiply 6 by 1000 to get a number with four numerals.

```
       1127
6000|6762    6000 can be subtracted from 6762 only        1
     6000
 600| 762    600 can be subtracted from 762 only          01
      600
 60|162   60 can be subtracted from 162 only              002
     120
 6|42     6 can be subtracted form 42 only                0007
   42                                                     1127 times
    0
```

    b. Let us divide 6762 by 6 again (review the definition of division).

```
       1127
 6 |6762    6 can be subtracted from 6762 only      1000 times
    6000
 6|762      6 can be subtracted from 762 only       0100 times
    600
 6|162      6 can be subtracted from 162 only       0020 times
    120
 6|42       6 can be subtracted from 42 only        0007 times
   42                                               1127
    0
```

c. Divide 1267 by 250.

$$\frac{5}{250\overline{)1267}}$$

$$\underline{1250}$$

17

250 can be subtracted from 1267   5 times

250 cannot be subtracted from 17 and becomes 17/250

$1267/250 = 5 + 17/250$

d. Let us divide $x^3 - 2x^2 - 17x + 10$ by $x - 5$. Note that $x - 5$ can be written as $x - 5 + 0 + 0$ without a change in value.

$$
\begin{array}{r}
0 + x^2 + 3x - 2 \\
x - 5 + 0 + 0\,\overline{)x^3 - 2x^2 - 17x + 10} \\
\underline{x^3 - 5x^2 + 0x + 0} \quad \text{(Review the definition of subtraction.)} \\
\underline{x - 5 + 0\,|\,0 + 3x^2 - 17x + 10} \\
3x^2 - 15x + 0 \\
\underline{x - 5\,|\,0 - 2x + 10} \\
\underline{|-2x + 10} \\
0 + 0
\end{array}
$$

Upon examination of the number line and powers, it can be seen that numbers such as 1001 can be written as powers of 10 since the number of alphabets of the language corresponds to the number called 10. Each part of the number can be represented as a power, and the number 1001 can be expressed as powers thusly:

$1001 = 1(10^3) + 0(10^2) + 0(10^1) + 1(10^0)$; that statement is true because $10^0 = 1$; $10^1 = 10$; $10^2 = 100$; $10^3 = 1000$.

By following the pattern above, $2x^3 + 1$ is actually in this case $2x^3 + 0(x^2) + 0(x^1) + 1(x^0)$, which can be written as $2x^3 + 0x^2 + 0x^1 + 1$.

Example: Divide $2x^3 + 1$ by $x + 2$.

$$
\begin{array}{r}
0 + 2x^2 - 4x + 12 + -23/(x - 5) \\
x + 2 + 0 + 0\,\overline{)2x^3 + 0x^2 + 0x^1 + 1} \\
\underline{2x^3 + 4x^2 + 0 + 0} \\
\underline{x + 2 + 0\,|\,0 - 4x^2 + 4x + 1} \\
0 - 4x^2 - 8x + 0 \\
\underline{x + 2\,|\,0 + 12x + 1} \\
12x + 24 \\
-23
\end{array}
$$

e. Divide $x^3 + x^2 - x - 10$ by $x - 2$

$$\begin{array}{r} 0 \ + \ x^2 + 3x + 5 \\ x - 2 + 0 + 0 \overline{|x^3 + x^2 - x - 10}} \\ x^3 - 2x^2 + 0 + 0 \\ x - 2 + 0 \overline{|0 + 3x^2 - x - 10}} \\ 0 + 3x^2 - 6x + 0 \\ x - 2 \overline{|0 + 5x - 10}} \\ 5x - 10 \\ 0 \ + 0 \end{array}$$

Divide 17 by 2

$$\begin{array}{r} 8 \\ 2\overline{|17}} \\ 16 \\ \hline 1 \end{array}$$

So, 2 can be subtracted from 17 8 times, but 2 cannot be subtracted from 1 because the group that contains two items is larger than the group that contains one item. Therefore, it cannot produce an integer. Closure necessitates that the outcome must be contained within the system of numbers and fractions are predicted by the process of division. The new numbers are parts of groups and are named *fractions* because it takes more than one to form a complete group. 1/2 is the new number created above and it is easy to see that it will take two of those (1/2) to make a whole group. It is seen that all fractions consist of two parts: a *numerator* (number of parts under consideration) and a *denominator* (number of parts contained within the group).

Decimals

Fractions can be written in a form called *decimals* by continuing the process of subtraction by separating the integral (whole) part from the fractional parts with a point of separation called a *decimal point* (1/2 is a fraction with the numerator smaller than the denominator). The subtraction process can be continued by writing the 1 (numerator) and a 0, which are separated by a decimal point as 1.0, and 2 is written as 0.2. (*remember that we can only add numbers that have the same name*). All fractions can be written in decimal form, and some are illustrated below.

	# of times 0.2 can be subtracted
1. $1.0 - 0.2 = 0.8$	1
2. $0.8 - 0.2 = 0.6$	2
3. $0.6 - 0.2 = 0.4$	3

4. 0.4 – .02 = 0.2   4
5. 0.2 – 0.2 = 0   5

0.2 can be subtracted from 1.0 a total of .05 times, and 1/2 = 0.5 as a decimal.

Example: Divide 17 by 5 (how many times can 5 be subtracted from 17?).

# of times 5 was subtracted

1. 17 – 5 = 12   1
2. 12 – 5 = 7   2
3. 7 – 5 = 2   3

# of times 0.5 was subtracted

1. 2.0 – 0.5 = 1.5   1
2. 1.5 – 0.5 = 1.0   2
3. 1.0 – 0.5 = 0.5   3
4. 0.5 – 0.5 = 0   4

Therefore 17/5 = 3 + 2/5 or 3.4 as a decimal.

Example: Divide 33 by 7.

# of times 7 was subtracted

1. 33 – 7 = 26   1
2. 26 – 7 = 19   2
3. 19 – 7 = 12   3
4. 12 – 7 = 5   4   33/7 = 4 + 5/7

Continue with the process to continue finding values.

# of times 0.7 was subtracted

1. 5.0 – 0.7 = 4.3   1
2. 4.3 – 0.7 = 3.6   2
3. 3.6 – 0.7 = 2.9   3
4. 2.9 – 0.7 =2.2   4

5. $2.2 - 0.7 = 1.5$ 5
6. $1.5 - 0.7 = 0.8$ 6
7. $0.8 - 0.7 = 0.1$ 7   (4 + 5/7 = 3.7 in decimal notation)

Continue with the process to continue finding values.

# of times 0.07 was subtracted

1. $.10 - .07 = .03$ 1   (4 + 5/7 = 3.71 in decimal notation)

Continue with the process to continue finding values

# of times 0.007 was subtracted

1. $.030 - .007 = .023$ 1
2. $.023 - .007 = .016$ 2
3. $.016 - .007 = .009$ 3
4. $.009 - .007 = .002$ 4   (4 + 5/7 = 4.714...)

The algorithm is less cumbersome, but this is an effort to show how the use of basic knowledge can be used to understand and arrive at valid conclusions.

Reciprocals

Reciprocals are numbers whose product is always positive 1.

     a.   Examples: $5(1/5) = 1$, $-5(1/-5) = 1$, $(3/4)(4/3) = 1$, $(-3/4)(4/-3) = 1$.

     b.   Uses of Reciprocals: Reciprocals can be used to change the appearance of a quantity without a correspondent change in its value. Reciprocals are very useful in many instances in daily commerce, including the sciences, engineering, and many other areas where projections and trends need to be considered.

     c.   If some basic information is known, then some transformations in appearance can be made using the power of reciprocals and the power of 1 as described above.

If it is known that $2x = 10$, the power of reciprocals can be used to find a solution.

$$2x = 10$$
$$2(1/2)x = 10(1/2)$$
Note that $2(1/2) = 1$ and $10(1/2) = 5$.
$$x = 5$$

Note that the transformation has been made using the power of reciprocals.

d. If one wishes to add $\frac{3}{4}$ and $\frac{1}{6}$, the use of reciprocals, allows one to transform the statements, and both will name the same quantity and can then be combined. $\frac{3(6)}{4(6)} + \frac{1(4)}{6(4)} = \frac{3(6)}{24} + \frac{1(4)}{24} = \frac{1(18+4)}{24} = \frac{22}{24}$

Note that each fraction was multiplied by the reciprocal of the other fraction. This can be transformed again by use of factoring and cancellation.

$$\frac{22}{24} = \frac{2(11)}{2(12)} = \frac{11}{12}$$

Note also that $\frac{6}{6}$ is the same quantity as $6(\frac{1}{6})$; $\frac{4}{4} = 4(\frac{1}{4})$.

That is a simple illustration of how what is known to be true can be used to discover additional truths.

Powers, Factors, and Exponents

When two or more parts of a number have the same name, they can be combined by using addition of powers, which is the same as adding numbers that name the same objects.

1. $b^2b^3 = b^{(2+3)} = b^5$
2. $(b^{2)3}(b^2)^4 = (b^2)^{(3+4)} = (b^2)^7 = b^{14}$
3. $(a+b)^2(a+b)^3 = (a+b)^{(2+3)} = (a+b)^5$

Note that the square of any two integers creates numbers called *perfect squares*.

a. Perfect squares are numbers that have two equal integral factors.
   1. $5(5)$, $6(6)$, $250(250)$, $9(9)$, $x(x)$, $2(2)$, $3(3)$…

b. When two or more parts of a fraction have the same name, they can be combined by using subtraction of powers, which is the same as subtracting numbers that name the same objects.

$$1. \frac{b^4}{b^3} = b^{(4-3)} \text{ or } \frac{1(b^4)}{1(b^3)} = \frac{1}{1b^{(3-4)}}$$

(Note that all numbers consist of two parts.)

c. When two or more parts of a fraction have the same name and are raised to the same power, they can be combined by using subtraction of powers, which is similar to subtracting numbers that name the same objects.

$$1. \frac{b^4}{b^4} = b^{(4-4)} = b^0 = 1 \text{ and } \frac{1(b^4)}{(b^4)} = \frac{1}{b^{(4-4)}} = \frac{1}{b^{(0)}} = \frac{1}{1} = 1$$

Note that any number raised to power of 0 has a value of 1.

Fractional and Negative Exponents

a. The *closure principle* dictates that all numbers can be used in operations involving numbers including exponents or raising to powers.
1. If x is raised to a power of 2 or x^2, then x^{-2} must also exist as a unique entity. It cannot equal x^2, which means x^{-2} is the reciprocal of x^2 and must be written as $\frac{1}{x^2}$.
2. If $x^{-2} = 1/x^2$, then $1/x^{-2}$ must equal x^2; that can be illustrated by using reciprocals to transform the statements into a different form or look as shown below.

$$1. \frac{1}{x^{-2}} = \frac{\frac{(1)(x^2)}{(1)(1)}}{\frac{(1)(x^2)}{(x^2)(1)}} = \frac{x^2}{1} = x^2 \text{ Note that } \frac{(1)(x^2)}{(x^2)(1)} = 1;$$

Example 1 shows that division is accomplished by multiplying by the reciprocal of the denominator with the use of the multiplicative identity.

$$2. \frac{a}{a} = a\left(\frac{1}{a}\right) = 1; \frac{a+b}{(a+b)} = (a+b)\frac{(1)}{(a+b)} = 1$$

$$3. \frac{\frac{a}{b}}{\frac{c}{d}} = \frac{a(d)}{b(c)} = \frac{ad}{bc}$$
$$\frac{c(d)}{d(c)}$$

Fractional Exponents as Roots of Numbers

a. If $2^0 = 1$ and $2^1 = 2$, it follows that $2^{1/2}$ must lie between 1 and 2 on the number line. Since the exponent (1/2) in $2^{1/2}$ is a fraction, we give it the name of *root* of a number.

$2^{1/2}$ is called the square root of 2.
$2^{1/3}$ is called the cube root of 2.
$2^{1/4}$ is called the fourth root of 2.
$2^{1/n}$ is called the nth root of 2.
$2^{2/n}$ is called the nth root of 2^2.
$2^{4/n}$ is called the nth root of 2^4.

Note that 2 is used as an illustration, but all numbers have roots (and some have roots that are integers).

    a.   Fractional powers obey the same rules as integral powers.

      1. 2^2 is read as the square of 2 or 2 squared or 2 to the second power.
      2. $2^{1/2}$ is read as the square root of 2.
      3. 2^3 is read as cube of 2 or 2 cubed or 3rd power.
      4. $2^{1/3}$ is read as the cube root of 2.
      5. 2^4 is read 2 to the 4th power.
      6. $2^{1/4}$ is read 4th root of 2.

    b.   Any *real number* can be raised to a fractional power.

      1. $2^{2/3}$ is read as the cube root of the square of 2.
      2. $2^{3/4}$ is read as the 4th root of the cube of 2.
      3. $(3/4)^{1/2} = \dfrac{3^{1/2}}{4^{1/2}}$
      4. $(2^{1/2})^{1/2} = 2^{(1/2)(1/2)} = 2^{1/4}$ (Review powers.)
      5. $[(2/3)^{1/2}]^{3/4} = (2/3)^{(1/2)(3/4)} = (2/3)^{3/8} = \dfrac{2^{3/8}}{3^{3/8}}$.

    c.   Study example carefully to determine why the statements are valid.
      1. Since $2^0 = 1$ and $2^1 = 2$, then $2^{1/2}$ must lie between 1 and 2 (or not be less than 1 or more than 2).

      2. If $4^2 = 16$ and $5^2 = 25$, and since 19 lies between 16 and 25, it follows that $19^{1/2}$ must lie between 4 and 5. $19^{1/2}$ is read as the square root of 19.

3. If $2^2 = 4$ and $3^3 = 9$ and since 6 is between 4 and 9, then $6^{1/2}$ must lie between 2 and 3 and is written as $2 < 6^{1/2} < 3$.

Numbers raised to a power of 1/2 are referred to as the square root of the number.

d. Fractional exponents sometimes create rational numbers that can be written as fractions.

    1. All rational numbers can be written as fractions (even if they are written using decimal notations). They can be recognized if there are repeating patterns in decimal notations or if they terminate.

    2. To write a repeating or terminating decimal as a fraction, multiply the original decimal portion of a number by a factor such that a section of the repeating decimal will be converted to a whole number part and a decimal part.

Example 1: If n = .98989898...., then multiplying n by 100 will produce 98.989898.... and the equation 100n = 98.9898...

Subtract 1n from 100 n to produce the statement:

100n = 98.9898...(means the digits repeat indefinitely)
1n = 00.9898
99n = 98 and n = 98; n is a rational number
        99

Example 2: If n =7.682682...., then multiplying n by 1000 will produce 7682.682682 and the equation 1000 n = 7682.682682...

1000n = 7682.682682....
1n = 0007.682682....
999n = 7675 and n = 7675/999

Example 3: If n= 0.333333...., then 10n = 3.33333...

10n = 3.33333....
1n = 0.33333...
9n = 3 and n = 3/9 or 1/3 (as a fraction)

e. Fractional exponents sometimes create irrational numbers that cannot be written as fractions.

$x = .5624352781$
$y = 2.341256987$
$2^{1/2}, 5^{1/2}, 7^{1/2}$ are irrational numbers. Use your calculator to computer their value and compare to x and y above. Compare your findings to the findings

in rational numbers and form your conclusion from the comparison of information.

Those are a few numbers that cannot be written as fractions. Can you find others?

Solving Equations

The definition of equations indicates probable methods of finding the value of quantities that will make statements true equations where the values of all quantities are known. Remember that in true equations, all values are known and can be identified on the basic number line. Variable equations contain some quantities that cannot be identified on the basic number line, and their identities must be discovered for them to become true equations.

a. To solve simple equations such as $3x = 27$, the solution is straightforward and can be solved with the use of reciprocals only.

$$3x = 27$$
$$1/3(3)x = 1/3(27)$$
$$1x = 9$$

b. To solve more complex equations such as $2x + 5 = 15$, the solution depends upon the manner in which you begin. You must always remember the conditions under which you are allowed perform an operation. (remember the definition of addition and subtraction)

1. You may begin by subtracting 5 from both members, or
2. You may begin by multiplying both members by the reciprocal of the coefficient of x.
   a. Either beginning will create a proper response as long as no untrue statements are inserted.

Solve $2x + 5 = 15$ by adding –5 to both members.
$2x + 5 – 5 = 15 – 5$ (Add – 5 to both members.)
$2x = 10$ (Simplify by combining like items.)
$1/2(2)x = 1/2(10)$ (Multiply both members by the reciprocal of 2.)
$x = 5$ (Simplified.)

Solve $2x + 5 = 15$ (Multiply both members by the reciprocal of 2.)
$1/2(2x + 5) = 1/2(15)$ (Multiply both members by the reciprocal of
$1/2(2x) + 1/2(5) = 1/2(15)$ (Use the distributive property.)
$x + 5/2 = 15/2$ (Simplify by use of reciprocals)
$x + 5/2 – 5/2 = 15/2 – 5/2$ (Subtract 5/2 from both members)
$x = 10/2$ (Simplify.)
$x = 5$

89

Note that if no facets of the system are violated, the solution depends only upon the method used to begin the solution. One method is preferred over the other method, but you are to choose the method best suited to your understanding of the stated conditions.

Translating Written Statements to Mathematical Statements

a. To translate written statements into mathematical statements:

1. Identify quantities and write out the description or name of the quantity.
2. Assign values to all identified values.
3. Compare statements of values to identical statements.
4. Write equation using the identical statements of values.

Example 1: John has $2.10 in his piggy bank. He knows that he has only put in nickels and dimes and that he only put twenty-seven coins into the bank. He wishes to know how many of each type of coin is in the bank.

This problem has two parts: one part dealing with the number of coins and the other part dealing with the amount of money in the bank. Remember that only quantities with the same named can be combined.

1. Identify
a. ? = number of nickels
b. ? = number of dimes
c. ? = total number of coins
d. 27 = total number of coins
e. ? = value of the nickels in the bank
f. ? = value of the dimes in the bank
g. ? = total value of the coins in the bank

2. Quantify
a. x = number of nickels
b. y = number of dimes
c. $x + y$ = total number of coins*
d. 27 = total number of coins*
e. $.05x$ = value of the nickels in the bank
f. $.10 y$ = value of the dimes in the bank
g. $.05x + 0.10y$ = total value of the coin in the bank**
h. 2.10 = total value of coins in the bank**

3. Compare (Analyze)

c. x + y = total number of coins*

d. 27 = total number of coins*

x + y = 27 is the equation for the number of coins in the bank, but neither the value of x or y is known—and one must be known before the other can be found. We must find another equation to find a value for either x or y with which the value of the other can be found.

g. .05x + 0.10y = total value of the coins in the bank **

h. 2.10 = total value of coins in the bank**

. 05x + 0.10 y = 2.10 is the second equation

Note that the problem is asking for the number of coins in the bank, and consequently all statements must refer to the number of coins in the first equation x + y = 27. The problem also required an accounting for the amount of money in the bank, and the second equation (.05x + 0.10 y = 2.10) is an accounting for the money. The equations are:

$$x + y = 27$$
$$05x + 0.10y = 2.10$$

Note that if the value of either x or y was known, then the value of the other can be found. x and y cannot be combined because they do not have the same name and cannot be added. In x + y =27, the value of x can be found by isolating it as follows:

$$x + y - y = 27 - y$$
$$x + 0 = 27 - y \text{ (by application of equation principles)}$$
$$x = 27 - y \text{ (x can be replaced in the other equation)}$$

In the equation .05x + 0.10 y = 2.10, replace x with 27 – y.

$$05(27 - y) + 0.10y = 2.10 \text{ (substitution)}$$
$$1.35 - .05y + 0.10 y = 2.10 \text{ (distributive property)}$$
$$1.35 + .05 y = 2.10 \text{ (addition)}$$
$$1.35 - 1.35 + .05y = 2.10 - 1.35 \text{ (equation property)}$$
$$.05y = 0.75$$
$$y = 15 \text{ coins (x + y = 27 becomes x + 15 = 27 and x = 12.)}$$

This is required by the definition of addition where all quantities to be added must have the same name. Therefore the number of nickels (27 – y) must be multiplied by the value of a nickel, and the number of dimes (y) must be multiplied by the value of a dime. Note that the solution requires your use of prior knowledge based upon your experience to help recognize the conditions for the equation to be written. The total amount represented by each coin is to be added and must be equal to the total amount in the bank. Since mathematics is a language and English is a language, the translation of math to English makes the solution much easier to decipher.

We have discovered that equations can be treated just as any other number and can be transformed by the operations of multiplication, division, addition, and subtraction as needed. The equation numbered 1 below may be multiplied by .05 and transformed by the equation principles (as shown below).

$x + y = 27$
$.05x + .05y = .05(27)$
$.05x + .05y = 1.35$ is the transformed equation number 1.

Now, solve the problem with the transformed equation numbered 1.

$$.05x + .05y = 1.35$$
$$\underline{.05x + .10y = 2.10}$$
$$\underline{-.05y = -.75}$$
$$-.05 \quad -.05$$
$$y = 15 \text{ and } x = 27 - 15 = 12$$

There are other possible solutions that may be limited only by your imagination and understanding of the principles of the basic system as previously outlined.

Example 2: Suppose you wanted to sell dinners for a charitable cause and had figured that your initial costs for food and supplies would be $900. If the labor cost to produce the dinners would be $0.50 per plate and you wished to sell the dinners for $8.00 each, how many dinners must be sold to break even with no loss and no gain?

1. Identify quantities to be useful in solving the problem.
a. ? = number of dinners needed to sell to break even at $8 per dinner
b. ? = amount of money obtained from sales
c. ? = labor cost to produce a dinner
d. ? = fixed expenses for each day

2. Assign values.

a. x = number of dinners needed to sell to break even at $8 per dinner

b. 8x = amount of money obtained from sales

c. 0.50x = labor cost to produce a dinner

d. 900 = fixed expenses for each day

3. Analyze.

a. $8x = amount of money needed to break even*

b. $0.50x = labor cost to produce the dinners needed to break even

c. $900 = fixed expenses for each day

d. $.50x + $900 = amount of money needed to break even*

The equation becomes $8x = 900 + .50x$.

Example 3: A lady had a basket of eggs for sale. She sold half of the eggs in the basket plus one half of an egg to a customer and had three eggs left in the basket. How many eggs were initially in the basket?

1. Identify

a. ? = initial number of eggs in the basket

b. ? = number of eggs she sold

c. ? = number of eggs remaining in the basket

d. ? = number of eggs initially in the basket

2. Assign values

a. x = initial number of eggs in basket *

b. $\frac{1}{2} x + \frac{1}{2}$ = number of eggs she sold

c. 3 = number of eggs remaining in the basket

d. $\frac{1}{2}x + \frac{1}{2} + 3$ = initial number of eggs in the basket *

3. Analyze

a. x = initial number of eggs in basket *

d. $\frac{1}{2}x + \frac{1}{2} + 3$ = initial number of eggs in basket *

The equation becomes $\frac{1}{2}x + \frac{1}{2} + 3 = x$.

The number of eggs sold plus the number remaining in the basket equals the number of eggs in the basket before the sale of the eggs.

Example 4: A backpack and the school supplies it contained cost $30.50. If the backpack cost $15.00 more than the supplies, how much did each item cost?

1. Identify (part 1 for cost of backpack and supplies)
a. ? = cost of backpack
b. ? = cost of the supplies
c. ? = cost of supplies and backpack
d. ? = cost of supplies and backpack
e. ? = cost of backpack – cost of the supplies
f. 15 = cost of backpack – cost of the supplies

2. Assign Values
a. x = cost of backpack
b. y = cost of the supplies
c. x + y = cost of backpack and supplies *
d. 30.50 = cost of backpack and supplies*
e. x – y = cost of backpack – cost of the supplies**
f. 15 = cost of the backpack – cost of supplies**

3. Analyze
c. x + y = cost of backpack and supplies *
d. 30.50 = cost of backpack and supplies*
The equation for part 1 is x + y = 30.50.
The value of either x or y must be found and substituted in the equation x + y = 30.50

Analyze (for part 2, cost of supplies)
e. x – y = cost of backpack – cost of supplies *
b. 15 = cost of backpack – cost of supplies*
x – y = 15 is the equation, and x = (15 + y) is the cost of the backpack. In the other equation, x + y = 30.50, replace the x with the value of x from the second equation to get (15 + y) + y = 30.50; y is the value of the supplies. Note that the equations had to be written with only one value represented to find a corresponding numerical value for the variable.
x – y = 15 solved for the numerical value of y show that y = 7.75 (cost of supplies).

Note that there are two equations, and you must choose which to use to help with the solution of the other equation. You must find a value for either x or y in one equation and replace that value into the other equation. By use of the equations from above, you will find x – y = 15, and solving for x, you will find a value x = 15 + y to substitute for the value of x in the other equation x + y = 30.50.

x + y = 30.50
(15 + y) + y = 30.50; substitute for (15 + y) for x
15 + 2y = 30.50
2y = 15.50
y = 7.75 and x = 22.75

The problem can also be solved by adding the equations as shown below.
1. x − y = 15
2. x + y = 30.50
   2x + 0 =45.50
x = 22.75 cost of backpack, and y = 7.75 cost of supplies.

Example 5: If working alone, one person can complete a job in five days and another can complete the same job in seven days. If you were to hire both to do the job, how many days can you expect it will take to complete the job?

1. Identify:
a. = time of work for the first worker to complete the job alone
b. = speed (rate) of first worker to complete the job alone
c. = time of work for the second worker to complete the job alone
d. = speed (rate) of second worker to complete the job alone
e. = time it took both to complete the job working together
f. = amount of work done by first worker
g. = amount of work done by second worker
h. 1 = job to be completed
i. = number of jobs to be completed

2. Assign Values
a. 5hrs = time of work for the first worker to complete the job alone
b. 1/5(hrs) = rate of first worker to complete the job alone
c. 7hrs = time of work for the second worker to complete the job alone
d. 1/7(hrs) = rate of second worker to complete the job alone
e. t = time it took both to finish the job
f. 1/5(t) = part of job completed by first worker*
g. 1/7(t) = part of job completed by second worker*
h. 1 = number of jobs to be completed *
i. 1/5(t) + 1/7(t) = number of jobs to be completed

Bobby Rabon

The equation to be solved is $\frac{1}{5}(t) + \frac{1}{7}(t) = 1$

Solving this problem—and others—requires the reader to use prior experiences or knowledge to find a solution. In the above problem, refer back to the meaning of the number line as movement along a sequence. It is clear that speed (rate) multiplied by time traveled will determine the distance traveled. Since the time of each worker is known and the number of jobs is known, the rate of the work can be determined, which implies that the amount (distance) of work is speed (rate) of work multiplied by the time it took to complete the work symbolized as d = rt. From the example above, it can be discerned that it took one worker five hours to complete one job by working at a certain speed (rate) and another worker took seven hours to complete one job. A relationship exists:

5hrs(r) = 1 (job) where r represents the speed of his work.
$7hrs(r_1) = 1$ (job), when r_1 represents the speed of his work.
The equation 5hr(r) = 1 can be solved by multiplying by the reciprocal of the coefficient

$\frac{5hrs(1)}{5hrs}(r) = \frac{1(1)}{5}$ and $r = \frac{1}{5}$ $\frac{7hr(1)}{7hr}(r_1) = \frac{1(1)}{7}$ and $r_1 = \frac{1}{7}$

The value of r means one worker did $\frac{1}{5}$ and the other worker did $\frac{1}{7}$ of the work.

The equation is the sum of the sum of the products of $rt + r_1t = 1$.

$\frac{1t}{5} + \frac{1t}{7} = 1$

Example 8: At a softball game, you use a stopwatch to measure the time it took for a pitch to travel 60 ft. to the batter. You wish to know how fast the pitch travels in miles per hour. If the pitch took .45 seconds to reach the plate, how fast was the pitch traveling in miles per hour?

60 feet = $\frac{(60 \text{ feet})(1 \text{ Mile})(60 \text{ minutes})(60 \text{ seconds})}{(.45 \text{ seconds})(5280 \text{ feet})(1 \text{ hour})(1 \text{ minute})}$ = $\frac{90.9 \text{ miles}}{1 \text{ hour}}$ or 90.9 miles per hour

Note that $\frac{1 \text{ mile}}{5280 \text{ ft.}} = 1$; $\frac{60 \text{ minutes}}{1 \text{ hour}} = 1$; $\frac{60 \text{ seconds}}{1 \text{ minute}} = 1$.

96

A merchant wishes to start a recycling program and wishes to instill a little mystery into the process by rebating the difference between the cost of the packaging material and the contents. The procedure he will use is as follows:

1. Rebates must be requested by the consumer.
    a. If the consumer requests an amount less than or equal to the cost of the container, the requested amount will be granted immediately.
    b. If the consumer requests an amount greater than the cost of the container, a second amount will be allowed. If the second amount is more than the cost of the container, the rebate will be forfeited.

Example: If the cost of a refrigerator and its container is $1150.00 and the cost of the refrigerator is $1100.00 more than the cost of the container, how much should the customer request as a rebate?

1. Identify and Assign Values
a. x = cost of refrigerator
b. y = cost of the container
c. 1150 = cost of refrigerator and the container
d. $1150 - y$ = cost of the refrigerator*
e. $1100 + y$ = cost of the refrigerator*

2. Analyze
d. $1150 - y$ = cost of refrigerator*
e. $1100 + y$ = cost of refrigerator*

The equation becomes $1150 - y = 1100 + y$
$$1150 - y + y - 1100 = 1100 - 1100 + 2y$$
$$50 = 2y$$
$$y = 25 \text{ (the cost of the container)}$$

Those examples are given for the reader to glean the diversity of issues that can be solved mathematically using the system developed thus far if the basis for the system is not compromised.

Dimensions, Functions, and Lines

There exist three spatial dimensions originating at the reference point "0" on the basic number line. Two of the lines create a flat (*plane*) surface of two dimensions. Inclusion of the third basic dimension creates a *space* that consists of three dimensions. At each point on a basic dimension, there are *dimensional lines* with the name of the point of origin on the basic dimension. An example is if a dimensional line originates at a point on the x dimension line called three, and another dimensional line originates at a point on the y dimension at point two, they will intersect (cross) at a point with the name of (3,2). The x dimension is always listed first, and the pair is called an ordered pair because (3,2) must be unique and is not the same as (2,3) which must also be unique. Several points are shown below. Compare and note the differences.

The diagram shows that points are named with two numerals or with a capital letter, and it takes two points to create a unique line as shown with the line labeled AB. Lines CD and EF illustrate a basic fact about two lines that intersect (cross) at exactly one point labeled G in the diagram. A careful examination of the lines will show that each value of x produces only one value for y. It can be seen that each value of x determines a unique value for y. Remember that unique means only one possible value.

y-axis

9							B		
8	E	2,8							
7							D		
6							7,7		
5	A				G				
4		3,4							
3							7,3		
2	5	3,2					F		
1			C	4,2					
0	1	2	3	4	5	6	7	8	9

The value of y is dependent on the value of x; x is named the *independent* variable, and y is called the *dependent* variable. y is also a function of x denoted by the expression y = f(x) because the value assigned to y depends on the value of x and there is only one value of y for each value of x. A line contains an unlimited number of points, which are unique to the line. Since a line contains points and points are represented as numbers that are unique and limitless, then a line is unique, contains limitless points, and has a unique name (illustrated as an equation). An equation of a line has the form of *ax + by = c (where a, b, and c are real numbers).*

An example of a unique line is $2x + 3y = 10$. No other line has that exact name. Since the value of y depends on the value of x, the value of y in the equation $y = f(x) = x + 5$ is found by substitution values of x in the equation $f(x) = x + 5$.
$f(2) = 2 + 5$ and $f(2) = 7$; when $x = 2$, then $y = f(2) = 7$.

Equations can be transformed into the proper format by application of equation principles.

Example 1: If $x + y = 5$ is the equation of a line, then $x - x + y = -x + 5$.
$y = -x + 5$ is the transformed equation.
Since $y = -x + 5$ and $y = f(x)$, then by the transitive property, we can write the equation as $f(x) = -x + 5$, and by computing the value of $f(x)$, we will have found the value of y in the original equation.

Example 2: If $-x + y = 5$, then $y = x + 5$, and $f(x) = x + 5$ because $y = f(x)$.

Find the value of $f(x)$.

$f(x) = x + 5$ when $x = 3$; $x = 2$; $x = 0$ and $x = -5$.
$f(3) = 3 + 5$ when $x = 3$, then $y = 8$ written as (3,8).
$f(0) = 0 + 5$ when $x = 0$, then $y = 5$ written as (0,5).
$f(-5) = -5 + 5$ when $x = -5$, then $y = 0$ written as (-5,0).
The points (3,8), (0,5), and (-5,0) are called coordinates and are always written in the form (x,y).

The diagram above also shows that two lines can cross (intersect) in only one point. Since lines are represented as equations that consist of points that are represented as numbers, lines and be added and subtracted or otherwise manipulated in the same fashion as numbers to find the common point of intersection.

Example 1: If $2x + 3y = 12$ intersects the line $2x - 3y = 4$, find the point of intersection.
$2x + 3y = 12$
$\underline{2x - 3y = 4}$ by adding the equations (the $+ 3y$ cancels the $- 3y$).
$4x + 0y = 16$ and $x = 4$; replace the value of x in either equation to find the corresponding value of y.

Example 2: $2x + 3y = 14$ intersects the line $2x + 7y = 22$.
$2x + 3y = 14$
$\underline{2x + 7y = 22}$ (by subtracting the equations—remember subtraction)
$0x - 4y = -8$ (by multiplying by the reciprocal of -4); $y = 2$ and $x = 8/3$

Example 3: $2x + 3y = 8$ intersects the line $4x + 5y = 20$; find the point of intersection.
1. To transform $2x + 3y = 8$, multiply by 2.
$$2(2x + 3y) = 2(8)$$
$4x + 6y = 16$ (Valid because both members were multiplied by 2.)

$4x + 6y = 16$ (The equations can now be paired.)
$\underline{4x + 5y = 20}$ (Subtract to eliminate x.)
$0 + 1y = -4$
Find x by using $4x + 5y = 20$.

$4x + 5(-4) = 20$; y replaced with -4.
$4x + 5(-4) = 20$.
$4x = 40$ (20 was added to each member.)
$x = 10$
The 2 lines have a common point at $(10,-4)$ and cross at that point.

Example: Find a solution for the variables in the two equations listed:

1. $2x - 2y = 12$
2. $4x - 5y = 9$

A possible solution can be as shown below:

Equation 1 can be solved for one of the variables:
$2x - 2y = 12$ can be solved for x as shown:
$2x - 2y + 2y = 12 + 2y$
$2x = 12 + 2y$
$x = \underline{12 + 2y}$ (Replace x in equation 2.)
2

Equation 2

$$4x - 5y = 9 \text{ becomes}$$
$$4\frac{(12 + 2y)}{2} -5y = 9$$
$$2(12 + 2y) - 5y = 9$$
$$24 + 4y - 5y = 9$$
$$24 - 24 - y = 9 - 24$$
$$1(-y) = -1(-15)$$
$$y = 15$$

That example could be solved with many different approaches including multiplication of each equation by a quantity that will allow a solution by addition or subtraction the to eliminate one of the variables. Inspections of the equations will show that adding or subtracting the equations will not eliminate one of the variables and will not allow for a solution. Equation principles allow for any operation to be performed on an equation if the same is done to both members. Consequently, the equations can be multiplied by any quantity if the principles of the system are not violated.

1. $2x - 2y = 12$ multiplied by 5 becomes $10x - 10y = 60$.
2. $4x - 5y = 9$ multiplied by 2 becomes $8x - 10y = 18$.

The equations can be subtracted to eliminate one of the variables and be solved in the traditional manner.

1. $10x - 10y = 60$
2. $\underline{8x - 10y = 18}$
   $2x = 42$; $x = 21$; $y = 15$

Bobby Rabon

If equation 2 had been multiplied by –2, it would have become –8x + 10y = –18, and the equations could have been added.

> 1. 10x – 10y = 60
> 2. –8x + 10y = –18
> 2x = 42; x = 21

These examples illustrate that solutions to problems can take many varied forms as envisioned by the solver. The only requirement is that no tenant of the system can be violated, and that is the reason why a careful scrutiny of the language of mathematics is imperative for a good grasp of the workings of basic mathematical functions and ideas.

A young artistic person decided to decorate some ornaments for sale as a fundraiser. She decided to decorate the ornaments and charge prices as shown below.

1. Plain ornaments sale for $1.00 each.
2. Additional ornaments will be sold at prices of $1.00 extra for each additional design.

She had ornaments with no deigns up to ornaments with a total of 12 designs.
One day, she sold one of each type of ornament and decided to tally up her sales for the day by adding up all of her sales slips of $1.00, $2.00, $3.00 … $12.00. She was late for an appointment and had to quickly determine the amount of the sales. How much were her sales for the day?

She knew that twelve items were sold, and she knew the value of each item. She decided to use that information to tally her sale and wrote:

> 1 Sales = 1 + 2 + 3 + 4 + 5 + 6 + 7 + 8 + 9 + 10 + 11 + 12
> 1 Sales = 12 + 11 + 10 + 9 + 8 + 7 + 6 + 5 + 4 + 3 + 2 + 1

She added the equations.

> 2 Sales = 13 + 13 +13 + 13 +13 +13 +13+ 13 + 13 + 13 + 13 + 13
> 2 Sales = 12(13)
> Sales = $\frac{156}{2}$ = $78.00

The next day, she sold one of each of the even-numbered ornaments.

Sales = 2 + 4 + 6 + 8 + 10 + 12
Sales = 12 + 10 + 8 + 6 + 4 + 2
2 sales = 14 + 14 + 14 + 14 + 14 + 14
2 sales = 6(14)
Sales = $\frac{84}{2}$ = $42.00

On the third day, she sold one each of the odd-numbered ornaments.
Sales = 1 + 3 + 5 + 7 + 9 + 11 for a total of 6 ornaments
Sales = 11 + 9 + 7 + 5 + 3 + 1
2 sales = 12 + 12 + 12 + 12 + 12 + 12
Sales = $36.00

How would you find the sum of the first 100 numbers? How about 300 numbers?

Notice that equations can be manipulated by using known information to formulate conditions that can be useful in solving problems. It attempts to illustrate that analysis of known information can lend to credible solutions to problems that may be difficult or time-consuming. The seeking of available information—sometimes given, but often implied—can lead to methods of finding solutions to problems that might otherwise be very difficult.

There are many other ideas that can be discussed here, but the focus is on getting an understanding of the structure of the system so that other ideas can be developed by the reader from the use of the basic ideas suggested by this text. A basic Algebra text will present many varied exercises that rely on basic comprehension and utilization of the principles of the system of mathematics. You must remember that any of the basic ideas of mathematics can be used concurrently and depend upon the insight of the problem solver.

Other ideas that will be developed require the development of additional ideas, but they all depend upon the basic fundamentals of the system. That is the reason why it is imperative that the reader makes a thorough and comprehensive study of the basic ideas and their uses.

Many rules expressed as shortcuts to solutions were developed by using many of the basic ideas of the system as well as other ideas developed from using the basic ideas created in the development of the system. The rules are called *algorithms* and can be used without a clear understanding of their origins or meanings. The algorithms make finding the answers much easier, but they often do not create an understanding of the fundamentals that are so vital to a thorough grounding in the concepts.

About the Author

The author has a master's degree, including thirty-plus additional hours in education. He was a math and science teacher, coach, and administrator for thirty-five years in Louisiana. As an administrator, he was in charge of student behavior and discipline, and he had the opportunity to interact with many students with various academic and behavioral skills. One very prominent facet of the interactions was that behavior is influenced by degrees of success—and mathematics was one of the main reasons for lack of academic success.

After retirement, he provided instruction in many subject areas in an alternative school for suspended and expelled students. Interactions with those students reaffirmed the idea that mathematics was a major problem and needed an injection of a new paradigm.

Printed in the United States
By Bookmasters